T0318793

HARNESSING THE POWER
OF VIRUSES

HARNESSING THE POWER OF VIRUSES

BORIANA MARINTCHEVA

Department of Biological Sciences, Bridgewater State University, Bridgewater, MA, USA

ACADEMIC PRESS

An imprint of Elsevier

Academic Press is an imprint of Elsevier
125 London Wall, London EC2Y 5AS, United Kingdom
525 B Street, Suite 1800, San Diego, CA 92101-4495, United States
50 Hampshire Street, 5th Floor, Cambridge, MA 02139, United States
The Boulevard, Langford Lane, Kidlington, Oxford OX5 1GB, United Kingdom

Library of Congress Cataloging-in-Publication Data
A catalog record for this book is available from the Library of Congress

British Library Cataloguing-in-Publication Data
A catalogue record for this book is available from the British Library

ISBN: 978-0-12-810514-6

For information on all Academic Press publications visit our website at
https://www.elsevier.com/books-and-journals

  Working together
to grow libraries in
developing countries

www.elsevier.com • www.bookaid.org

Publisher: John Fedor
Acquisition Editor: Linda Versteeg-Buschman
Editorial Project Manager: Fenton Coulthurst
Production Project Manager: Mohanapriyan Rajendran
Designer: Greg Harris

Typeset by TNQ Books and Journals

Contents

Biography

Dr. Boriana Marintcheva is an Associate Professor of Biological Sciences at Bridgewater State University, Bridgewater, Massachusetts, where she teaches Virology, Cell Biology, and Molecular Biology courses with laboratories. By training, Dr. Marintcheva is a molecular virologist who has studied *Herpes Simplex Virus Type I* replication as a doctoral student at the University of Connecticut Health Center and the biology of the bacteriophage *T7* single-stranded DNA–binding protein as a postdoctoral fellow at Harvard Medical School. While completing her BS/MS degree in Biochemistry and Microbiology at Sofia University, Bulgaria, she briefly investigated stress response in plants. Dr. Marintcheva is passionate about science education, science promotion, and development of pedagogical tools advancing teaching and learning. She views technology (in all shapes and forms) as a potential driver of curiosity, creativity, and motivation to learn, as well as powerful evidence for the positive impact of science on human life. She believes that it never hurts to be optimistic, even when viruses are involved, and is perpetually fascinated by the turns and twists they bring to science and technology.

Acknowledgments

To everyone who fosters curiosity and creativity and considers the implications of his/her findings beyond the limits of his/her current project!

I am grateful for the support of my home institution, Bridgewater State University, for giving me the gift of time to focus on this work as a Presidential Fellow for the 2016–17 academic year. Special thanks to President Clark, the Center for Advancement of Research and Scholarship, Andy Holman, editor of The Bridgewater Review, Maureen Vernon from the Dean's office of the College of Science and Math, and my colleagues in the Biology Department. Many thanks to my graduate advisor, Dr. Sandra Weller, University of Connecticut Health Center and my postdoctoral advisor, Dr. Charles Richardson, Harvard Medical School, for their support over the years. Thanks to my family and friends for riding the roller coaster of life by my side.

This book refers to the work of countless scientists throughout the existence of virology as a discipline. I extend my deepest appreciation to everyone who has contributed to understanding viruses and regret that it is impossible to mention every valuable piece of work and to reference specifically every discovery.

Introduction to Viral Structure, Diversity and Biology*

Tiny, deadly, fascinating. Viruses have been around us for thousands of years and have impacted our society regardless how well we understood them. It is hard to imagine that when the word virus is mentioned, something positive will come to mind. After all, viruses got their name from the Latin word for poison, which fits them perfectly when picturing the devastating diseases they cause in humans, animals, and plants. When computers came around and self-replicating programs became the fact of life, the term virus gained a new meaning: again not exactly a positive one. Not that long time ago the Internet brought to us the idea of viral videos. Today we can describe as viral not only meningitis (hoping that we will never need to deal with one) but also anything funny, crazy, amazing, or spectacular caught on video. Now thanks to "YouTube", the idea of Latin word for poison being associated with something different from disaster does not sound absolutely outrageous, does it?

1.1 WHAT ARE VIRUSES, ANYWAY?

Viruses are tiny nonliving agents with very complicated lives. Figuratively, they have been best described as "a piece of bad news wrapped in a protein" by Sir Peter Medawar, a Nobel laureate and father of organ transplantation. Viruses are built from nucleic acid packed in

*Parts of this chapter were originally published in Marintcheva B. A box of paradoxes: the fascinating world of viruses. *Bridgew Rev* 2013;**32**(2):25–8. http://vc.bridgew.edu/br_rev/vol32/iss2/8 and are reproduced here with the permission of the editor.

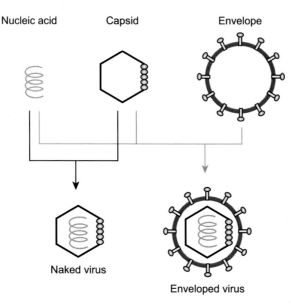

FIGURE 1.1   **Structural components of viruses.** Depending on their structural elements viruses in nature fall in two large groups: naked and enveloped viruses. Naked viruses consist of viral genome/nucleic acid (DNA or RNA) and protein-based viral capsid, collectively forming a nucleocapsid. Enveloped viruses have an additional membrane-based layer called an envelope.

protective protein coat (a capsid), and some are wrapped in an additional layer of lipids and proteins (an envelope), thus some viruses are nonenveloped or naked (yes, this is an official scientific term for viruses composed from nucleic acid and capsid only), and others are enveloped (Fig. 1.1). Viruses are known to man for less than 150 years. They were first described late in the 19th century when scientists were searching for the infectious agent causing discoloration and necrotic spots on tobacco leaves. It so happened that the relevant pest, tobacco mosaic virus (TMV), was not retained on bacterial filters and thus a new "poison," a virus, became known to science. The discovery was a collective result of the independent work of several scientists: Adolf Mayer, who established that the leaf discoloration can be transmitted by juices extracted from diseased plant; Dmitrii Iwanowski (Ivanovsky) and Martinus Beijerick, who realized that the infectious agent passes through filters retaining bacteria. Of course the latter would not be possible without the development of clay bacterial filters by Chamberland in 1884. For a long time the concept of what actually constitutes a virus was very mundane. Different bits of evidence and ideas were consistent with viruses being fluid (passing through bacterial filters); particles (plaques assays), or molecules (chemical composition assays, sedimentation analysis). It took roughly half of a century to actually become possible to visualize a virus by crystallization or electron microscopy, and much longer to understand how viruses are built and how they operate. In 1957, Lwoff put together an article reflecting on current ideas (at that time) on the nature of viruses only to write "viruses are viruses." The golden era of molecular biology contributed to understanding how viruses reproduce and allowed for integration of classical virology knowledge and the molecular basis of viral characteristics. The studies of animal and human viruses benefited from the discovery that many of them can propagate outside the host organism in chicken eggs and in tissue culture. All of us living in the 21st century have the tremendous privilege to learn about viruses in a rather structured way and awe at the intellectual effort that has allowed the contemporary understanding of what viruses are.

Despite the huge diversity in the viral world, one can define viruses based on hallmark features relevant to their structure and interactions with the host. Any virus can be described as a small nonliving infectious agent that is able to productively propagate in a specific host cell. Each virus contains either DNA or RNA genome coding for viral components and molecules guiding the execution of productive viral infection. As nonliving agents, viruses are passive entities whose fate is completely dependent on the surrounding environment, regardless what our need of interactive expression makes them to do. Viruses themselves do not strategize, exploit, take advantage of, and so on. We, as humans, describe them in many ways similar to our human experiences while trying to rationalize how they work. In reality, any virus outside the host cell is completely inert. Viruses propagate only in cells that provide appropriate environment and virus–host interactions are a function of evolution.

Figuratively, one can think about viruses as boxes of paradoxes. They are considered nonliving agents; however, being "dead" does not prevent them from causing disease or executing various life styles. Another paradox is the huge disproportion between the virus size and the magnitude of complexity they govern. Physically, viruses measure on the nanometer scale. The flu virus measures 80–120 nm in diameter, therefore if we line up flu particles along a cross section of a human hair, we will fit about 1000 of them (Fig. 1.2). On the molecular level, viruses could be considered as "kings and queens" of exceptions, "utilizing" unique strategies for genome replication and synthesis of viral components. That makes them an unparalleled resource for molecular tools that can be applied toward new technology development. So, how are viruses built anyway?

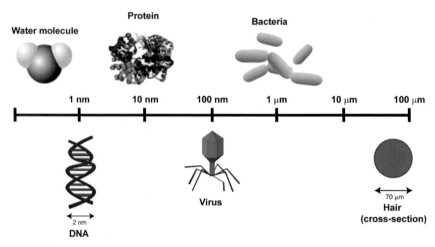

FIGURE 1.2 **The scale of viruses.** Most viruses measure on the nanometer scale and are readily observable only with an electron microscope. One exception is giruses, such as Mimivirus, which measure on micrometer scale and can be observed under conventional light microscope. *This image was originally published in Marintcheva B. A box of paradoxes: the fascinating world of viruses.* Bridgew Rev *(a campus faculty magazine) 2013;***32***(2):25–8.* http://vc.bridgew.edu/br_rev/vol32/iss2/8 *and is reproduced with the permission of the editor.*

1.2 OVERVIEW OF VIRAL STRUCTURE: CAPSIDS, GENOMES, AND ENVELOPES

1.2.1 Capsids

Scientists characterize viruses by systematically evaluating the components they are built of and tracking the interactions of the viral particles with the host, both on cellular and organismal level. Naked viruses are comprised of genome and protein-based capsid. In contrast, enveloped viruses have an extra lipid-based layer, called an envelope (Fig. 1.1). Morphologically, viruses are described based on their shape as helical, isometric, and complex (Fig. 1.3). The overall shape of naked viruses is determined by the structure of their capsid. Classical example of helical virus is TMV, which appears as a thin cylinder in electron micrographs. One can figuratively think about TMV as corn on the cob in which the RNA genome builds the cob, whereas the kennels collectively represent the protein capsid, built from repetitive units or capsomers. Helical viruses are sometimes also called filamentous. For example, the shapes of Ebola virus and bacteriophage M13 are often described as filamentous. Isometric viruses are highly symmetrical and geometrically resemble icosahedron, a structure with 20 faces. Isometric capsids are built very much like soccer balls from repetitive capsomers with defined shape. If a soccer ball was indeed a viral capsid, it would contain two types of capsomers: white

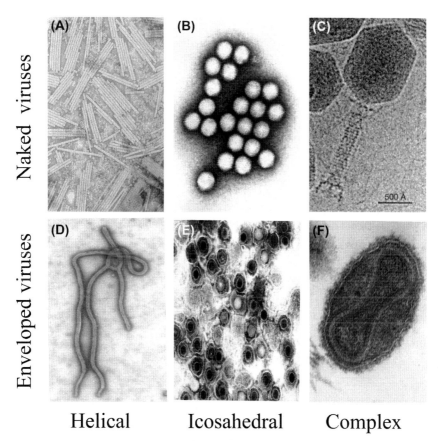

FIGURE 1.3 **Electron micrographs of naked and enveloped viruses with various shapes.** Examples of naked viruses and enveloped viruses are displayed in the top and bottom rows, respectively. Examples of viruses with helical shape are shown on the left, icosahedral viruses are shown in the middle, and viruses with complex shape are shown on the right. (A) Tobacco mosaic virus. (B) Adenovirus. (C) Bacteriophage T4. (D) Ebola virus. (E) Herpes simplex virus type 1. (F) Smallpox virus. *(A) Image courtesy of T. Moravec. (B) Image courtesy of Dr. G. William Gary, Jr./Center for Disease Control (CDC). (C) Reprinted from Rossmann MG, Mesyanzhinov VV, Arisaka F, Leiman PG. The bacteriophage T4 DNA injection machine.* Curr Opin Struct Biol *2004;**14**(2):171–80, with permission from Elsevier. (D) Image courtesy of Cynthia Goldsmith/CDC. (E) Image courtesy of Dr. Fred Murphy/CDC. (F) Image courtesy of Dr. Fred Murphy and Sylvia Whitfield/CDC.*

hexagons and black pentagons. The shapes of the capsomers and the symmetry of icosahedrons vary significantly from virus to virus (Fig. 1.4). Adenoviruses, polioviruses, herpesviruses, and papilloma viruses, for example, are all isometric viruses. The flu virus presents an interesting paradigm in terms of viral shape. Although its genome fragments are packed in helical nucleocapsids, the overall shape of the virus is spherical (similar to isometric viruses) due to the presence of

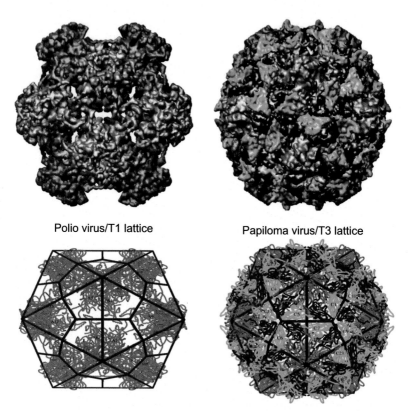

Polio virus/T1 lattice   Papiloma virus/T3 lattice

FIGURE 1.4 **Icosahedral viral capsids with different symmetry.** 3D reconstruction of human papilloma virus (HPV) capsid with T1 lattice (on the left) and poliovirus capsid with T3 lattice. The HPV capsid is built from one type of a capsomer, a penton, composed of five copies of the major capsid protein L1 (the minor capsid protein L2 is not shown). Recombinant L1 protein can self-assemble in a stable capsid in vitro and is used as a vaccine. The polio virus capsid is built from two capsomers: a penton in blue and a hexon in green/red. *The images were generated using VIPERdb software* (http://viperdb.scripps.edu; *Carrillo-Tripp M, Shepherd CM, Borelli IA, Venkataraman S, Lander G, Natarajan P, Johnson JE, Brooks 3rd CL, Reddy VS. VIPERdb2: an enhanced and web API enabled relational database for structural virology.* Nucleic Acids Res 2009;**37**(Database issue):D436–42.) *and pdb files* **1DZL** (Wien MW, Curry S, Filman DJ, Hogle JM. Structural studies of poliovirus mutants that overcome receptor defects. Nat Struct Biol 1997;**4**(8):666–74.) and **1ASJ** (Chen XS, Garcea RL, Goldberg I, Casini G, Harrison SC. Structure of small virus-like particles assembled from the L1 protein of human papillomavirus 16. Mol Cell 2000;5(3):557–67).

the viral envelope surrounding all eight genomic fragments (Fig. 1.5). Viruses with complex shape are extremely diverse group with bacteriophage T4 "leading the pack" as the most frequent example of textbook image of a virus. T4 has well distinguishable head, tail, and attachment fibers, which in general are not typical for viruses. Viral capsids can be viewed as architectural masterpieces assembled with minimal number

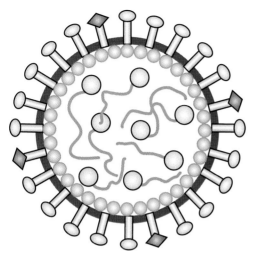

FIGURE 1.5 **Structure of the influenza virus.** Influenza A virus is an enveloped (−) ssRNA virus with 8-piece segmented genome. Each segment of RNA (*orange line*) is encapsulated in individual helical capsid (not drawn for simplicity) and carries an attached RNA polymerase (*grey circle*). All eight helical nucleocapsids are wrapped in an envelope (*black circle*), supported by a layer of matrix proteins (*orange circles*). Surface antigens hemagglutinin (*grey ovals*) and neuroniminidase (*grey rhomboids*) protrude from the outer surface of the envelope. Despite the helical symmetry of the nucleocapsids, the overall shape of the virion is close to spherical.

of building protein blocks. For example, the capsid of the polio virus is built from only three types of proteins (Fig. 1.4) elegantly assembling a robust icosahedral structure. Molecules present on the outside surface of naked viruses are instrumental for establishing productive contact with the host cell and essentially work as ligands or binding partners for the host's viral receptors.

1.2.2 Genomes

Viral genomes consist of DNA or RNA only, never both. DNA and RNA molecules can be double stranded or single stranded, linear or circular (Fig. 1.6), segmented (composed of multiple pieces of nucleic acid) or nonsegmented. The meaning of the term genomic segment could be confusing as its use is driven by both nomenclature principles and historic reasons. Strictly speaking, a genome segment is an individual and unique piece of nucleic acid among multiple pieces comprising one whole viral genome. For example, the influenza A virus has segmented genome comprised of eight ssRNA segments (Fig. 1.5). Herpesviruses, which have nonsegmented genomes composed of one linear dsDNA molecule, have the so-called UL (unique long) and US (unique short) segments, corresponding to regions/portions of their genomes flanked by repeats. The HIV genome can be somewhat confusing too with each virion carrying two copies of the same ssRNA molecule. The HIV genome is considered nonsegmented, and the two ssRNA molecules are called copies not segments. In many viruses the genome ends contain repeated sequences, chemical modifications, or secondary structures, which often have regulatory functions. Genomes

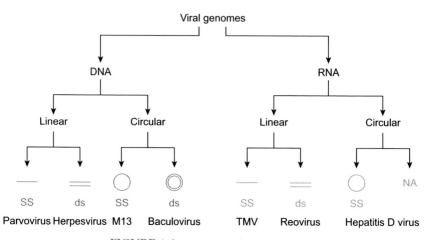

FIGURE 1.6   Diversity of viral genomes.

are tightly packed inside the capsids and frequently the genome and the capsid are collectively called nucleocapsid. Amazingly, viruses are able to execute productive infection and of course make us sick with very limited genetic information. The flu virus genome, for example, contains only 15,000 nucleotides. For comparison the human genome is 3,200,000,000 nucleotides or approximately 200,000 times longer. Needless to say, viruses have to be superefficient, in their quest to invade the host cell and to propagate. Bacteriophage Qβ is among the smallest RNA viruses with a genome built from 4217 nucleotides and only 4 genes. Among the smallest known animal DNA viruses is TT virus whose genome is comprised of less than 4000 nucleotides and 4 predicted genes. On the opposite side of the scale is the giant *Megavirus chilensis* with genome as large as 1.3 MB coding for 1000 genes. What functions viral genomes code for is a tantalizing question. Analysis of viral genome sequences revealed that approximately 80% of the viral genomes code for virus-specific genes, many of which have no known homologues or known function.

1.2.3 Envelopes

Viral envelopes are built from phospholipid bilayers that resemble cellular membrane. The lipid components are purely of host origin, whereas the protein components are virally coded. Most of the envelope proteins are integral glycoproteins with critical function for virus–host interactions. Similar to the proteins on the outside surface of naked viruses, they function as ligands to host receptors. Enveloped viruses acquire their envelopes as they are leaving the host cell.

1.2.4 Additional Virion Components

Some enveloped viruses have distinctive protein layer spanning the space between the capsid and the envelope. In herpesviruses, this layer is named tegument, whereas in HIV and influenza a matrix (Fig. 1.5). It is thought that the extra layer is stabilizing the virion filling the gap of non-existing direct connections between the nucleocapsid and the envelope. A small number of extra molecules are found in the virions (viral particles) of many viruses. These include viral enzymes, transcription activators, molecules assisting with nucleic acid packing such as polyamines, and some cellular molecules. HIV is a great example in this respect. Its virion contains copies of reverse transcriptase and integrase, both essential enzymes for fast and efficient onset of genome replications. Amazingly, a host tRNA used as a primer for the reverse transcriptase is also found in the HIV particle.

1.3 VIRUS DIVERSITY AND CLASSIFICATION: ORIGIN OF VIRUSES

Viruses are everywhere and have been described to infect organisms from every single kingdom. Probably, every species on the planet is infected by viruses; however, so far science has managed to study mainly the ones that are relevant to human health. It has been estimated that 10^4–10^8 bacteriophages exist per milliliter of water in the Earth's aquatic system, 10^9 bacteriophages inhabitate each gram of soil and that probably at least 320,000 mammalian viruses are yet to be discovered. Regardless of the limitations of the estimation approaches, it is straightforward to conclude that our planet harbors huge number of viruses and thus viruses vastly outnumber any group of organisms. Similar to living organisms, viruses are a product of evolution and are classified in attempt to better describe and study them. The International Committee on Taxonomy of Viruses (ICTV) is the organization in charge of developing and updating universal taxonomy classification. Viruses are classified in orders (with suffix -virales), families (-viridae), subfamilies (-virinae), genuses (-virus), and species. The latest report of the ICTV (as of 2016) lists 2284 viruses and viroids grouped in 6 orders, 87 families, 19 subfamilies, and 349 genera. Due to the large number of not very well-characterized viruses, the classification system is very far from complete. The Baltimore classification groups viruses according to the chemical nature of their genomes and their mode of propagation, specifically focusing on mechanisms of genome replication and generation of mRNA compatible with the host (Table 1.1). Viral genomes are described by their chemical nature (DNA vs. RNA; double stranded vs. single stranded) and polarity. Sense/positive

TABLE 1.1   Baltimore Classification of Viruses

Class	Type of Nucleic Acid	Example
Class I	dsDNA	Herpesvirus
Class II	ssDNA	Bacteriophage M13
Class III	dsRNA	Reovirus
Class IV	(+) ssRNA	Poliovirus
Class V	(−) ssRNA	Influenza virus
Class VI	RNA (with RT)	HIV
Class VII	DNA (with RT)	Hepatitis B virus

Baltimore classification groups viruses based on the chemical composition and structure of their genome, as well as their replication strategies.
ds, double stranded; *RT*, reverse transcription; *ss*, single stranded; *(−)*, negative strand; *(+)*, positive strand.

sense/(+) genomes consist of DNA or RNA strand that is identical to the mRNA sequence (except for T vs. U diferences in DNA and RNA). Antisense/negative sense/(−) genomes consist of DNA or RNA strands complementary to the corresponding mRNA. Initially, the Baltimore system was proposed in 1971 and then included six classes grouping only animal viruses. Class VII, dedicated to dsDNA viruses that utilize reverse transcription was added later with hepatitis B virus being the most studied example. Viruses infecting organisms from all kingdoms are currently classified according to the Baltimore system.

The discovery of giant dsDNA Mimivirus led to the introduction of the term girus as a designation of giant viruses (Fig. 1.7). Amazingly, giant viruses code for some of the components of cellular translational machinery with the extreme case of Klosneuviruses encoding for extended set of translation players including all 20 aminoacyl-tRNA synthetases. The discovery of Sputnik, an entity colonizing Mamavirus virions and interfering with their virulence, gave rise of virophages as designation of viruses "infecting" viruses. At this moment, it is controversial if virophages are just a type of satellite viruses or a truly novel entity. Satellite viruses are satellites that code for their own capsid protein(s). Satellites are subviral agents (DNA or RNA based) that replicate only in the context of coinfection with a "helper" virus coding for indispensable functions in trans. Satellites are to be distinguished by viroids, which are the smallest described pathogens. Viroids are "naked" ssRNA agents found only in plants that do not code any proteins and do not require any known functions coded in trans for their propagation.

How and when viruses came to existence are burning questions ultimately connected to origins and evolution of living organisms. As the

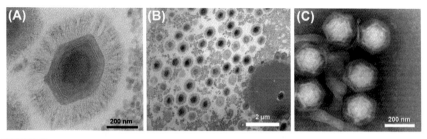

FIGURE 1.7   **Electron micrographs of giruses and virophages.** Mimivirus was the first discovered girus measuring 750 nm in diameter, including the pili. When replicating, giruses set up replication factories in the host cell allowing for efficient process. Mimivirus replication could be slowed down if the host cell is coinfected with virophages. Virophages are subviral agents that can propagate only in cells coinfected with cognate/helper virus. (A) Mimivirus (scale 200 nm); (B) Mimivirus viral factory in its amoeba host (scale 1 μm); (C) Virophages (scale 200 nm). *Image courtesy Aherfi S, Colson P, La Scola B, Raoult D. Giant viruses of Amoebas: an update.* Front Microbiol *2016;7(349).* https://doi.org/10.3389/fmicb.2016.00349, *CC-BY,* https://creativecommons.org/licenses/by/4.0/.

number of sequenced viral genomes increases, detailed studies regarding their evolution are emerging and shedding new light on the viral world and its connections to the rest of life on the planet. Two major hypotheses are currently being discussed, which simplistically can be summarized as virus-first or cell-first hypothesis. Virus-first hypothesis postulates that viruses came to existence independently before or alongside cells in the primordial RNA world and were instrumental shaping life on Earth. Presumably, RNA viruses existed alongside RNA genome-based ancient cells, and eventually some of them evolved to DNA viruses. DNA viruses infecting host cells with RNA genomes are considered drivers for cellular evolution resulting in the rise of modern cells with DNA-based genomes, which formed the three current domains of life. The cell-first hypothesis postulates that cells existed on the planet before viruses and viruses "escaped" from them gaining ability to "move" to cells close to their original "home," i.e., bacterial viruses are escapees from bacterial cells, archaeal viruses are escapees of archaeal cells and so on. An alternative version of the cell-first hypothesis proposes that viruses were cells once and have gradually regressed by losing major components and thus their ability to reproduce/exist independently. Detailed arguments in favor or against each hypothesis can be found in the specialized literature. For the purpose of the discussion here, it will be sufficient to mention that the discovery of the transposons is the strongest evidence in favor of the "escapee" origins of viruses, whereas the discovery of the giruses and the presence of genes coding for components of the cellular translational machinery are the major supporting evidence for the regression mechanism of virus emergence. An extreme version of the regression version of cell-first hypothesis is the idea about prior existence of cells of fourth domain of life that regressed and gave rise of viruses. The recent advancements in the field of giruses resulted in revisiting the definition of viruses and life and the idea of capsid-based organisms (viruses) and ribosome-based organisms (cells). Viral genomics is very fast advancing field that is expected to provide further clues on viral origins and evolution. Recent research efforts elegantly demonstrated distinct prokaryotic roots of both RNA and DNA eukaryotic viruses. A diagram of the evolution of large dsDNA viruses is shown on Fig. 1.8 as a representative example.

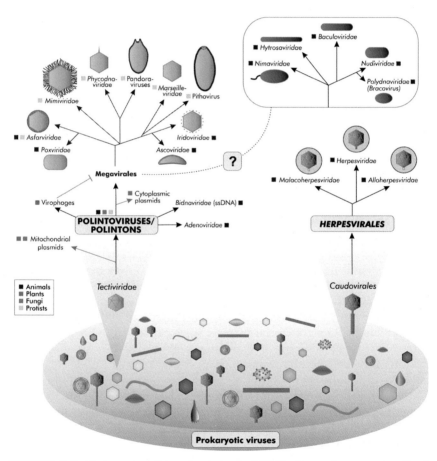

FIGURE 1.8 Prokaryotic origins of eukaryotic dsDNA viruses. *Image courtesy Koonin EV, Dolja VV, Krupovic M. Origins and evolution of viruses of eukaryotes: the ultimate modularity.* Virology 2015;**479–480**:2–25; https://doi.org/10.1016/j.virol.2015.02.039, *CC-BY*, https://creativecommons.org/licenses/by/4.0/.

1.4 APPROACHES TO VISUALIZE VIRUSES

Due to their size, viruses are impossible to observe directly without specialized equipment. Electron microscopy is the ultimate technique to visualize viruses. Despite its resolution the approach offers information about limited number of virus characteristics (shape of the virion and capsid, overall shape of surface molecules, and unusual genome structures) and even less means for following viruses around. Science usually tracks viruses by their effects on the host. For example, the discoloration of tobacco leaves is used as a tool to follow TMV infection. Cold sores are a painful visual for herpes simplex virus type 1 (HSV-1) and so on. In the context of laboratory experiments, viruses are tracked and quantified using plaque assays. Viral plaques present themselves as areas clear of cellular growth in a lawn of bacterial cells or monolayer of eukaryotic cells following viral infection. Functionally, each plaque can be defined as the progeny of one single plaque forming unit/virus. Plaque assays were initially developed to study bacteriophages. A mixture of the virus in question, the relevant host cells, and melted soft agar is overlaid on the surface of regular agar plate. When left undisturbed the soft agar solidifies forming a uniform thin layer, which practically immobilizes the bacteria and the virus particles in the mixture. The division of the uninfected cells results in a bacterial lawn with milky appearance. Infected cells are destroyed and release newly generated virions, which in turn infect neighboring cells. After several rounds of viral propagation, areas free of bacterial growth appear as clear spots in the milky bacterial lawn (Fig. 1.9). Plaques of eukaryotic viruses can be observed if a monolayer of permissive cell line is infected with the virus of interest and the cells are overlaid with semisolid material, such as methylcellulose or agarose, limiting the diffusion of newly released virions. Plaques are visualized by staining with vital dyes (crystal violet or methyl red) that color only living cells. Areas in which cells have been lysed by the virus appear colorless, defining individual plaques. Plaques can be observed easily under microscope, especially if the infecting virus expresses fluorescent protein such as green fluorescent protein (GFP). If plaques are counted, conclusions can be made for the viral titer (number of viruses per unit volume) of the initial sample. Plaque sizes and morphology are used to derive information aiding virus identification or key characteristics of the infection. It is important to realize that plaques are "man-made phenomenon" and do not really exist in nature. If a viral infection takes place in liquid culture where cells move, areas free of cell growth cannot be observed. Instead, the bacterial/eukaryotic cell culture gradually decreases in density as the infection proceeds until all cells are lysed. Plaque assays became technically possible for animal and plant viruses once culturing cells in monolayers was established in the mid-20th century, several decades after plaque assays were established for bacteriophages. This is a great example how advances from

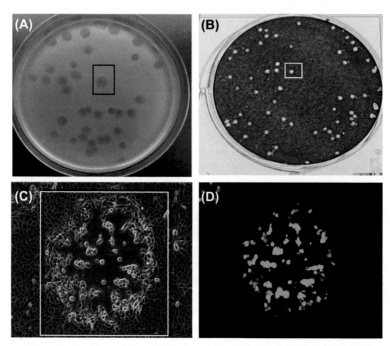

FIGURE 1.9 **Viral plaques.** (A) Phage plaques on bacterial lawn; (B) Monkeypox plaques on a lawn of Vero E6 cells stained with crystal violet; (C) African green monkey BS-C-1 cells infected with EGFP expressing vaccinia virus WR strain bright field microscopy image, (D) fluorescent microscopy image of panel C. Single plaques are boxed with rectangles in panels A, B, and C. *(A) Image courtesy Colom J, Cano-Sarabia M, Otero J, Aríñez-Soriano J, Cortés P, Maspoch D, Llagostera M. Microencapsulation with alginate/CaCO(3): A strategy for improved phage therapy.* Sci Rep *2017;7:41441.* https://doi.org/10.1038/srep41441, *CC-BY,* http://creativecommons.org/licenses/by/4.0/. *(B) Image courtesy Mucker EM, Chapman J, Huzella LM, Huggins JW, Shamblin J, Robinson CG, Hensley LE. Susceptibility of Marmosets* (Callithrix jacchus) *to Monkeypox virus: a low dose prospective model for Monkeypox and Smallpox disease.* PLoS One *2015;10(7):e0131742.* https://doi.org/10.1371/journal.pone.0131742, *CC-BY,* https://creativecommons.org/licenses/by/4.0/. *(C and D) Images courtesy of Sherry L. Haller and Stefan Rothenburg, Kansas State University;* http://www.k-state.edu/media/newsreleases/sept15/rothenburg9915.html.

different scientific fields enhance each other and allow new tools for virus research to be created.

1.5 VIRAL LIFE CYCLES: LYTIC VERSUS LATENT

Although viruses are nonliving agents, they have complex life styles, and some are able to switch between two known alternatives: lytic and latent life cycle. Lytic viruses are direct killers. They infect their host, propagate, and destroy the cell allowing the release of newly made

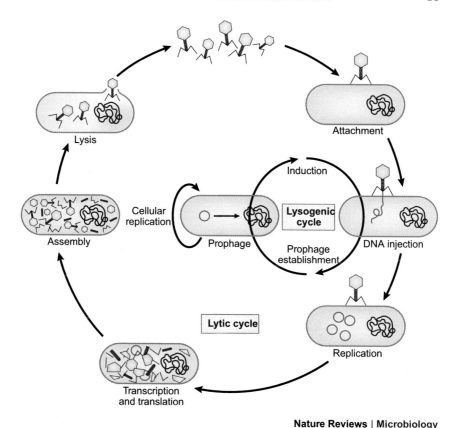

Attachment

Lysis

Induction

Cellular replication

Lysogenic cycle

DNA injection

Assembly

Prophage

Prophage establishment

Lytic cycle

Replication

Transcription and translation

Nature Reviews | Microbiology

FIGURE 1.10 **Life cycle of bacteriophage lambda.** Bacteriophage lambda is a classic example of a temperate phage, which can undergo lytic or lysogenic/latent life cycle. During the lytic life cycle the phage propagates extensively, and numerous new viruses are released to the environment, whereas during lysogeny, lambda exists as a prophage (provirus in general) integrated into the host genome, and its DNA is replicated only as a function of the host replication without release of new virions *Reprinted from Salmond GP, Fineran PC. A century of the phage: past, present and future.* Nat Rev Microbiol 2015;*13(12):777–86, with permission from Macmillan Publishers Ltd.*

viral particles. Naked viruses generally lyse the infected cell directly, whereas most enveloped viruses destroy the cell by stealing membrane material. Eventually, the rate of virion release outpaces the rate of membrane production, and the cellular membrane is depleted, thus no longer supporting the integrity of the cell. Latent viruses tend to exist quietly in their hosts for a long time and occasionally reactivate producing bursts of new viruses. Reactivation is essentially a switch between latent and lytic life cycle, which allows a latent virus to execute productive infection and new virions to be produced (Fig. 1.10). The viral genome can physically integrate in the host genome (bacteriophage lambda; HIV) or persist in the host cytoplasm as a free episome (HSV-1, bacteriophage P22). If the host cell is dividing the viral genome is passively replicated, i.e., as a "natural" part of the cellular genome without production of viral particles, every time cellular division occurs. The metabolic activity of latent viruses is very limited and extends only to viral products relevant to maintenance of the latent state. In general, latency and reactivation are well studied in very few viruses. In lambda bacteriophage, they are associated with an interplay of transcription repressors and proteins modulating their activity. Herpes simplex virus type 1 (HSV-1), the causative agent of cold sores, provides a good everyday life illustration how lytic and latent life styles work together (Fig. 1.11). Most people in the world are infected with HSV-1 early in life. The infection manifests itself as a cold sore, which is practically a bunch of blisters full of viruses. While the immune system is generally effective in taking care of the viruses circulating in the body, some viruses become permanent residents of the dorsal root ganglia and establish dormancy or latent infection. When the body is experiencing high levels of stress (big deadline, intense UV exposure on the beach, or severe cold) the virus could reactivate and travel back to the lip, where it executes lytic infection resulting in a new cold sore. Latency and reactivation in HSV-1 are mediated by RNA transcripts coded by the LAT (latency-associated) region of the viral genome.

Typical lytic life cycle proceeds through several steps: adsorption/attachment; penetration/viral entry; biosynthesis; assembly; and release.

FIGURE 1.11 **Life cycle of herpes simplex virus type 1.** HSV-1 enters epithelial cells via membrane fusion, travels along the cytoskeleton highways to the nucleus. Uncoating takes at the nuclear pore releasing the naked viral genome in the nucleus. HSV-1 genes are transcribed sequentially, a feature common for dsDNA viruses, in three distinct classes: (1) immediate early (IE) genes coding for functions associated with gene expression and establishment of productive infection; (2) early (E) genes coding for replication proteins; and (3) late (L) genes, coding for capsid and envelope proteins. VP16, an HSV-1 tegument protein, and the cellular factor, Oct1, are instrumental for the onset of gene expression. Once viral proteins and genomes are synthesized, the HSV-1 capsid is assembled in the nucleus and buds through the nuclear membrane acquiring a temporary envelope. It is thought that the herpes nucleocapsid travels across the endomembrane system to reach the limits of the cells where it exits (viral egress) acquiring its final envelope. The released viruses can infect the nerve endings in the vicinity of the site of primary infection, where they can establish lytic or latent infection. The choice of life cycle is at least in part driven by VP16 by not well-understood mechanism. Once latency is established, it is maintained by neuronal factors (Oct2) viral transcripts from the LAT (latency) locus. HSV-1 can be reactivated as a consequence of body or environmental triggers, and the newly generated virions travel back to the initial site of infection giving us a new cold sore. *Reprinted from Simonato M, Manservigi R, Marconi P, Glorioso J. Gene transfer into neurones for the molecular analysis of behaviour: focus on herpes simplex vectors.* Trends Neurosci 2000;**23**(5);183–90, *with permission from Elsevier.*

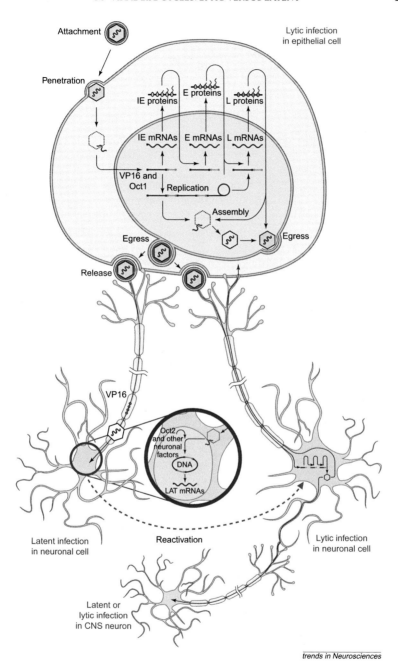

Specific details are discussed on the examples of bacteriophage lambda (Fig. 1.10) and herpes simplex virus type 1 (Fig. 1.11) focusing on the big picture and highlighting conceptual differences between prokaryotic and eukaryotic viruses.

1.5.1 Adsorption/Attachment

To infect the host cell viruses need to attach to its surface in a productive way. The process is mediated by interaction between molecules residing on the outer surface of the virus (viral ligands) and the host cell (cellular receptors), respectively. The interactions are dependent on complementarity of shapes and biochemical characteristics of cellular receptors and viral ligands, as well as the environmental conditions. To initiate infection, a virus has to stably and specifically attach to the cell surface for sufficient time, so that subsequent molecular events can be triggered. Proteins and glycoproteins from the capsid of the naked viruses and the envelope of the enveloped viruses are the molecules that serve as viral ligands in the process of attachments. Proteins, glycoproteins and polysaccharides serve as receptors on the host side. In some cases, receptors are aided by coreceptors, which by themselves cannot initiate the virus–host interaction; however, can drastically improve its efficiency once the initial contact is established. A great example in this regard is HIV, which uses CD4 molecule as a receptor and CCR5 or CXCR4 molecules as coreceptors. Individuals lacking functional CCR5 molecules are resistant to HIV infection, whereas heterozygous individuals are "slow progresses" and experience HIV-related symptoms after longer latent period. Needless to state, the above specifics are explored as possible gateways to design anti-HIV therapeutic approaches.

1.5.2 Penetration/Viral Entry

At this step, viral genomes physically "pass" across the structures (cell wall, cellular membrane, nuclear membrane) separating them from the cellular environment supporting their propagation. Since bacterial, animal, and plant cells are built differently the viruses infecting them have evolved very different strategies to enter the host cell.

Once stably attached to the cellular surface via receptor-mediated adsorption, bacterial viruses inject their genomes through the cell wall and cellular membrane. The exact mechanism how the process works and what are the driving forces of genome entry are not well understood and subject of intense discussion between biologists and physicist. Most likely, tailed and nontailed bacteriophages employ slightly different strategies; however, from the perspective of viral life cycles they accomplish the same end point: placement of the viral genome in environment conducive for its expression and replication. Bacteriophage penetration is the biological process that allowed Hershey and Chase to perform their brilliant experiments providing direct evidence in support of DNA as the molecular basis of heredity.

Animal viruses do not need to deal with hard cell walls; however, they need to cross the cellular membrane, the underlying cellular cortex and to navigate the internal complexity of the cell to reach the area permissive for their propagation. Many viruses can only propagate in the nucleus, thus their genomes have to cross the nuclear membrane too. Enveloped animal viruses enter the cell via membrane fusion or endocytosis. In the first case, the physical entity entering the cytoplasm is the nucleocapsid, whereas the envelope becomes part of the cellular membrane. In the second case the entire viral particle (nucleocapsid and envelope) enter the cytoplasm as internal content of the generated endocytic vesicle. Naked viruses enter the cell also via endocytosis. The described modes of entry are receptor-mediated and allow viruses to penetrate the cell using preexisting cellular pathways for moving substances across the cellular membrane.

In order viral genomes to be expressed and replicated, they need to be freed of their "packaging" material or uncoated. Most frequently, virus capsids are disintegrated in the cytoplasm with the help of the lysosomes. Viruses replicating in the nucleus reach their destination using the transport networks of the cytoskeleton. In some cases, the nucleocapsids dock to the nuclear pores and unload the genome in the nucleus (HSV-1); in others the entire nucleocapsid enters the nucleus and the uncoating takes place there (influenza, hepatitis B).

Plant viruses face completely different challenges entering their host cells. The plant cell wall is thick, hard, and sturdy, thus the strategies described for bacterial and animal viruses are not feasible. Plant viruses penetrate the cell following mechanical injury or are transmitted by insect vectors. Once propagating in the plant, viruses can cause systemic infections spreading through the plasmodesmata and the phloem transport system of the plant. Plasmodesmata are microchannels/cytoplasm bridges connecting neighboring plant cells and allowing transfer of substances and signals between them. Many plant viruses are known to code for viral proteins that interact with the plasmodesmata and modify them to allow virus passage.

1.5.3 Biosynthesis

Once the viral genome finds itself in an environment supporting its expression and copying, viral components are made in vast amounts on a rapid time scale. Essentially, the virus genetic program guides cellular physiology and metabolism toward preferential synthesis of viral components, which are quickly assembled into new viral particles. Naturally, the enormous diversity of the virus world translates in diverse mechanisms of taking over the cell and using cellular machinery to fulfill viral functions. Nevertheless, the processes of synthesis of viral DNA, RNA, and proteins are comprised of limited (not small but limited) number of steps and employ various players and regulatory elements that can be targeted to interfere with the viral infection or put to work as molecular tools. In essence, viral genomes have to remain intact for sufficient time to allow genome expression and copying, and the host cell has to be maintained alive while energy and building blocks for viral molecules are being utilized for biosynthetic purposes. Productive biosynthesis results in generation of many copies of the viral genome, capsid, and envelope proteins (if applicable) to be assembled into new virions. Fig. 1.12 depicts the key characteristics of the replication process for all seven classes of viruses according to the Baltimore classification. Three types of enzyme activities carry on the physical polymerization of viral genomes: DNA-dependent DNA polymerases (DdDps), which replicate the vast majority of ssDNA and dsDNA viruses using the conventional central dogma replication mechanism of synthesizing a new DNA strand on a template of preexisting one. The majority of RNA viruses are replicated by RNA-dependent

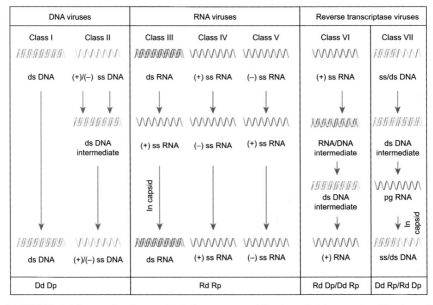

FIGURE 1.12 **Mechanisms of viral genome replication.** Baltimore classification of viruses and their modes of replication and generation of viral mRNA. DNA genomes are depicted with *black helices*, whereas RNA genomes are depicted with *orange helices*. The enzymes replicating each class genome are abbreviated as follows: *DdDp*, DNA-dependent DNA polymerase; *RdDp*, RNA-dependent DNA polymerase or reverse transcriptase; *RdRp*, RNA-dependent RNA polymerase.

RNA polymerases (RdRps), which synthesize new RNA molecules using preexisting ones as templates. dsRNA viruses execute their replication in rather unusual way. Initially, only copies of the sense RNA strand are generated. In parallel, capsid proteins are made and capsids assembled. Each capsid is populated with one sense RNA strand with attached polymerase, which synthesizes the antisense RNA strand completing the dsRNA genome. The described mechanism is driven by evolution. Since dsRNA is a molecule *nongrata* in the cell, the "in capsid" completion of the dsRNA replication presents a clever hiding trick bypassing the cellular antiviral defenses.

Viruses using reverse transcription also have elaborated replication mechanisms. Retroviruses (Class VI of Baltimore classification with best studied example HIV) are RNA viruses, whose ssRNA genome is converted into RNA/DNA hybrid by the polymerase activity of the enzyme reverse transcriptase. At the next step, the RNAse H activity of the enzyme degrades the RNA component of the RNA/DNA hybrids opening the stage for the polymerase activity to synthesize the second DNA strand resulting in a dsDNA version of the HIV genome. The dsDNA genome is then integrated into the host DNA and replicated on the DNA level with every round of cellular division. HIV viral genomes are generated from the dsDNA copy of HIV inserted into the host genome by cellular RNA polymerase II, i.e., the enzyme that transcribes cellular mRNAs. Regardless of the rather complex mechanism of replication, HIV propagates very fast and is able to do so, at least in part, due to the fact that it packs two genome copies in its capsid, each coming with preattached molecules of reverse transcriptase and integrase enzymes that ensure immediate execution of the reverse transcription steps and fast integration in the host genome.

The replication of DNA viruses that use reverse transcription is not well understood at this moment. The best studied virus of this class (Class VII) is hepatitis B virus. Its genome is rather interesting. It is neither double stranded nor single stranded. It is circular, but none of the DNA strands is a closed circle. The answer to the riddle comes down to a gapped dsDNA genome made up of two partially complementary ssDNA strands. When hepatitis B infects its host the viral genome is transported to the nucleus where it is converted in a closed dsDNA molecule, also called covalent closed circle DNA or cccDNA. cccDNA is used to generate mRNAs and pgRNA (pregenomic RNA). pgRNA is packaged into a preformed capsid together with a molecule of the viral reverse transcriptase. The latter generates the gapped dsDNA genome in multistep process using its reverse trancriptase, RNAse H and DNA polymerase activities. HIV and hepatitis B viruses use reverse transcription for very different purposes. HIV utilizes it as a tool to "transform" its genome into a form consistent with the host genome, so that it can be integrated into cellular DNA. In contrast, hepatitis B uses reverse transcription to generate a rather unconventional genome inside its capsid.

Viral mRNAs of DNA viruses are synthesized by conventional transcription employing DNA-dependent RNA polymerase (DdRp) that utilizes dsDNA template of the viral genome itself (Class I), dsDNA intermediate (Class II), integrated dsDNA genome copy (Class VI) or cccDNA (Class VII) (Fig. 1.13). Some viruses are transcribed with virally coded enzymes; others take advantage of the cellular ones. Classes III and IV generate mRNA with the action of virally coded RNA-dependent RNA polymerases. Viral mRNA is translated by the cellular translational machinery.

Several conceptual strategies for productive propagation have been repeatedly described in the virus world: (1) many viruses code for key genome-copying enzymes such as DNA and RNA polymerases, primases, helicases; (2) many viruses code for enzymes ensuring abundant pool of nucleotides to support genome copying. For example, HSV-1 (a dsDNA virus) codes for ribonucleotide reductase, an enzyme that converts ribonucleotides in deoxyribonucleotides, thus supplementing the host cell enzyme with the same function and eliminating a possible rate-limiting step/bottleneck for fast genome replication. Bacteriophage T7 codes for potent exonuclease and endonuclease, which are thought to be instrumental in degrading the host genome to nucleotides subsequently used for de novo synthesis of viral genomes; (3) many viruses employ strategies

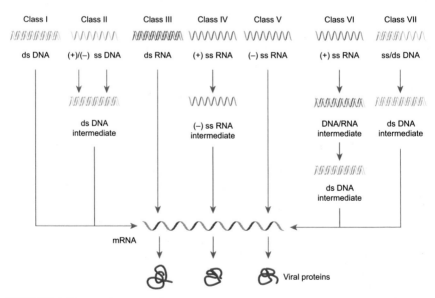

FIGURE 1.13   **Mechanisms of viral mRNA synthesis.** Baltimore classification of viruses and their mechanisms of mRNA generation. DNA is depicted with *black helices*, whereas RNA is depicted with *orange helices*.

to ensure preferential translation of viral proteins. For example, the flu virus snatches CAPs (CAP is a nucleotide-based structure at the 5′ end of mRNAs essential for initiation of protein synthesis) from the host mRNAs thus preventing the ribosomes from translating them. In the same time the snatched CAPs are being used to "decorate" the flu mRNAs making them compatible with the ribosomes. One can easily compare viruses to very creative and resourceful spies that manage to highjack cellular components, introduce switches in the cell physiology, and force the host to produce viral building blocks instead of normal cellular parts and molecules. Among the most fascinating viral tricks are the ones allowing many viral functions to be coded with limited genome size such as alternative reading frames, alternative translation initiation codons, and ribosome frameshifting.

1.5.4 Assembly

The process of putting together new viral particle is extremely complex and hardly well understood for most viruses. The specifics of the process are dependent on the type of the virus (naked vs. enveloped), type of capsid, location of the assembly process in the cell and the complexity of the viral particle. It is one type of task to put together a naked virus with nonsegmented genome, and completely different task to put together an enveloped virus with segmented genome. Regardless of complexity, all viruses are metastable: they have to be "sturdy enough" to survive in the environment and "fragile enough" to fall apart during infection, so that they can easily propagate. Some viruses spontaneously self-assemble once critical concentration of viral components is reached; others are assisted by specialized viral and cellular proteins. As a rule of thumb, the capsids of helical viruses self-assemble on the genome driven by protein–protein interactions between the capsomers and protein–nucleic acid interactions between the genome and the capsomers. TMV can self-assemble in a test tube if purified capsid protein and viral RNA are mixed together. TMV capsid protein can even self-associate in a helical structure by itself in absence of viral RNA. Some icosahedral capsids can self-assemble around the viral genome; others do so via complex pathways involving the assistance of nonstructural viral proteins, called scaffold proteins and/or cellular chaperones. Scaffold proteins are present only in intermediate steps of the process and released when the capsid shell adopts stable shape. For example, bacteriophage T4 and herpesviruses assemble their capsids with the help of virally coded scaffold proteins (Fig. 1.14). When they are released the internal volume of the structure is populated by the viral genome, in other words, the genome is packaged in a preformed capsid. In bacteriophage T4 the assembled nucleocapsid serves as a platform to assemble the tail and the attachment fibers resulting into a mature virion.

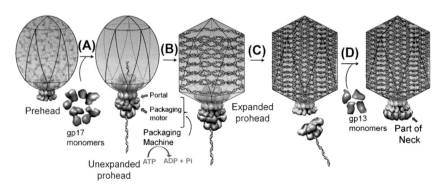

FIGURE 1.14   **Virion assembly of bacteriophage T4.** Virion assembly is understood the best in dsDNA viruses that assemble by essentially the same sequence of steps with variation in the specific details. First, capsid proteins assemble a prehead with the help of scaffolding proteins. Scaffold proteins are then removed proteolytically to make room for the upcoming genomes (A), whereas the capsid proteins undergo conformational change to form expanded prohead (B). DNA genomes are packaged (B,C) through a single capsid vertex called a portal, which is equipped with a DNA packaging and processing machinery involving a DNA motor (gp17 [gene product 17]). Once DNA is packaged, the DNA packaging machinery is replaced by neck components (gp13), which function as a platform for the tail assembly (D). The mature virion has an angular head full of DNA, neck, and tail with attachment fibers (not shown). The portal structure is colored in changing color (grey, blue, green, and yellow) to emphasize conformational changes that take place at different steps. Designations of capsid intermediates may vary between different viruses. For example, in herpesviruses intermediates are labeled by consecutive letters of the alphabet (A,B,C); however, the general principle of assembly is the same suggesting evolutionary conservation of the process. *Image courtesy Zhang Z, Kottadiel VI, Vafabakhsh R, Dai L, Chemla YR, Ha T, Rao VB. A promiscuous DNA packaging machine from bacteriophage T4.* PLoS Biol *2011;9 (2):e1000592.* https://doi.org/10.1371/journal.pbio.1000592, *CC-BY,* https://creativecommons.org/licenses/by/4.0/.

In herpes, the assembled nucleocapsid leaves the nucleus and travels through the guts of the endomembrane system to reach the cellular membrane where it acquires its envelope and it is released to the environment. In some cases, capsids undergo an additional maturation step involving proteolytic processing of capsid proteins, which can happen even after the envelope is acquired. The viral genomes can be packaged by length or by sequence. Most viruses use elaborate protein machinery and nucleic acid signal to package the viral genome; others would package any piece of nucleic acid that happens to be around depending solely on length. The latter strategy can be used in the lab to package nucleic acid of interest and employ virus particles as a delivery vehicles. The packaging of genomes in preassembled capsids takes place using one single entry point equipped with molecular motor, called a capsid portal. Once, nucleocapsids of the naked viruses are assembled, they are ready to be released. Enveloped viruses acquire their envelopes budding through the cellular membrane or through the endomembrane system of the eukaryotic cell. Needless to say, there is a huge diversity in approaches and mechanisms how viruses assemble their nucleocapsids and acquire envelopes. Generally, the more components are involved, the more sophisticated the process is. Viruses replicating in the nucleus have more complex assembly than the ones replicating in the cytoplasm.

1.5.5 Release

The release of newly produced virions is associated with instant lysis (naked viruses) or continuous release of viral particles via budding through the cellular membrane (enveloped viruses). The mechanisms of release are directly dependent on the host cell and the structures limiting its borders. In bacterial systems, viruses have to break through cellular membrane (one membrane in gram-positive bacteria and two membranes in gram-negative bacteria) and cell wall composed of peptoglycan. Most frequently the job is accomplished by the highly coordinated action of holins, spanins (proteins making holes in the bacterial membrane[s]) and endolysins (proteins degrading the cell wall). Few viruses deal with the cell wall by coding for inhibitors of cell wall synthesis. In plants, viruses have to overcome a very thick and sturdy polysasscharide-based cell wall. Generally, plant viruses take advantage of the plasmodesmata and phloem to spread throughout the entire plant or leave cells through places of injury inflicted by environmental conditions and insects. In order to exit animal cells viruses need to break the cellular membrane. In many cases the membrane breaks by itself unable to hold the vast amount of naked viruses that were just assembled in the cell; in other cases, naked viruses utilize viroporins or lytic lipids to controllably lyse the cellular membrane. Enveloped viruses break the membrane by pinching out small areas to build their envelopes. Released viruses are ready to infect permissive cells in their surroundings and to propagate again. If no appropriate host is around, they persist for a while depending on their stability and the specific characteristics of the environment and eventually disintegrate.

The viral world is incredibly diverse and science is only starting to understand it and appreciate its impact on the life of our planet. Only a limited number of viruses have been studied in detail allowing scientists to tell detailed stories of their lives as nonliving agents in the world of cells. Increasing number of viral components and tricks are continuously being used to generate more knowledge about living organisms and help solving practical problems in our society. In a way, viruses are becoming "viral" in their own way to fascinate us even further.

Further Reading

1. Koonin EV, Dolja VV. Virus world as an evolutionary network of viruses and capsidless selfish elements. *Microbiol Mol Biol Rev* 2014;**78**(2):278–303.
2. Schulz F, Yutin N, Ivanova NN, Ortega DR, Lee TK, Vierheilig J, Daims H, Horn M, Wagner M, Jensen GJ, Kyrpides NC, Koonin EV, Woyke T. Giant viruses with an expanded complement of translation system components. *Science* 2017;**356**(6333):82–5.
3. Buchmann JP, Holmes EC. Cell walls and the convergent evolution of the viral envelope. *Microbiol Mol Biol Rev* 2015;**79**(4):403–18.
4. Raoult D, Forterre P. Redefining viruses: lessons from Mimivirus. *Nat Rev Microbiol* 2008;**6**(4):315–9.
5. Desnues C, Boyer M, Raoult D. Chapter 3-Sputnik, a virophage infecting the viral domain of life. In: Małgorzata Ł, Wacław TS, editors. Małgorzata Ł, Wacław TS, editors. *Advances in virus research*, vol. 82. Academic Press; 2012. p. 63–89.
6. Summers WC. Inventing viruses. *Annu Rev Virol* 2014;**1**(1):25–35.
7. Gonzalez ME, Carrasco L. Viroporins. *FEBS Lett* 2003;**552**(1):28–34.
8. Smith AE, Helenius A. How viruses enter animal cells. *Science* 2004;**304**(5668):237–42.
9. Baltimore D. Expression of animal virus genomes. *Bacteriol Rev* 1971;**35**(3):235–41.
10. Anthony SJ, Epstein JH, Murray KA, Navarrete-Macias I, Zambrana-Torrelio CM, Solovyov A, Ojeda-Flores R, Arrigo NC, Islam A, Ali Khan S, Hosseini P, Bogich TL, Olival KJ, Sanchez-Leon MD, Karesh WB, Goldstein T, Luby SP, Morse SS, Mazet JAK, Daszak P, Lipkin WI. A strategy to estimate unknown viral diversity in mammals. *mBio* 2013;**4**(5). e00598–13.

Viral Tools for In Vitro Manipulations of Nucleic Acids: Molecular Cloning

Without a doubt, molecular biology is the biological subdiscipline, where the beneficial impact of viruses is easiest to identify and appreciate. The majority of the initial knowledge regarding the significance of DNA as the molecule containing the genetic code of living matter, and the processes involved in the flow of genetic information was derived from experimentation with bacteriophages. *Escherichia coli* and the bacteriophages infecting it were the sources of the first tools in what is now an extensive array of molecular biology techniques and approaches allowing for diverse modes of manipulations of nucleic acids in vitro.

2.1 OVERVIEW OF MOLECULAR CLONING

2.1.1 Structure and Chemical Composition of Nucleic Acids: Flow of Genetic Information

DNA and RNA are linear biopolymers built from nucleotides connected with phosphodiester bonds (Fig. 2.1). Chemically, each nucleotide is composed of nitrogen base (A, G, C, T, and U), sugar (deoxyribose or ribose), and a phosphate. By convention the positions of atoms in the nitrogen base are labeled with consecutive Arabic numbers, whereas the position of atoms in the sugar component are designated with Arabic numbers followed by the prime symbol, i.e., number 3 refers to the third atom in the base, and number $3'$ refers to the third atom in the sugar moiety of each nucleotide. The base is always attached to the $1'$ position of the sugar via glycosidic bond, and the phosphate is always attached in $5'$ position via ester bond. Deoxyribose, the sugar component in DNA, and ribose, the sugar component in RNA, differ in one functional group. Ribose has hydroxyl group in

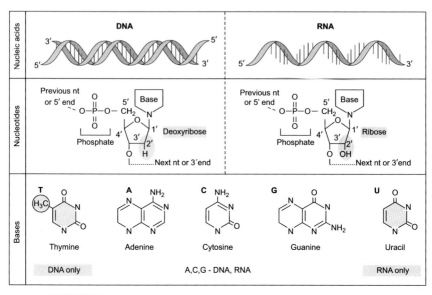

FIGURE 2.1   Nucleic acid's chemical composition and basic structure.

position 2′, whereas deoxyribose has only one hydrogen in that position. Nucleotides are connected with phosphodiester bridge formed by a dehydration reaction joining the 5′ phosphate and 3′ hydroxyl of adjacent monomers. The chemical structure of nucleic acids determines their polarity. One of the ends of the polynucleotide chain has a free 5′ phosphate group, and it is designated as the 5′ end or the beginning of the molecule. The other end has a free 3′ hydroxyl group and is designated as the 3′ end of the molecule. The above convention is universally used to describe directionalities in processes involving nucleic acids as well as to list DNA/RNA sequences. If the 5′ and 3′ designations are not included, it is assumed that the 5′ end is on the left and the 3′ is on the right. The sugar and the phosphate comprise the negatively charged backbone of all nucleic acid molecules, whereas their unique properties are function of the specific order of nucleotides (as specified by their base) along the length of the molecule. DNA molecules contain adenine (A), thymine (T), guanine (G), and cytosine (C) bases. In RNA, T is replaced with uracil (U). The complementarity in chemical structure and their ability to form hydrogen bonds allows nucleotides to form particular base pairs according to the complementarity rule. G and C base pair together forming three hydrogen bonds. A base-pairs with T or U, forming two hydrogen bonds. Each base pair contains one purine (two ring base; G or A) and one pyrimidine base (C or T/U), resulting in a highly regular polynucleotide structure. The two strands in the DNA double helix are antiparallel as the 5′ end of each strand is positioned across the 3′ end of the other strand. Although single stranded, RNA molecules have complex secondary structure composed of partially double-stranded regions (hairpins) and single stranded regions (loops). The secondary structure of both nucleic acids plays key role in their biological function. The genetic code is based on triplets of nucleotides specifying for amino acids. The genetic code is redundant and degenerated composed of 1 start codon (AUG, coding for the amino acid methionine), 3 stop codons specifying end of protein synthesis, and 60 codons specifying the remaining 19 amino acids, some of which are coded by multiple codons.

Cellular genomes are defined as the sum of all DNA in the cell and include chromosomal and plasmid DNA in prokaryotes and chromosomal DNA, organelle DNA (mitochondria and chloroplast) and plasmids (generally found in single-cellular eukaryotes and plants) in eukaryotes. Plasmids are small circular molecules of DNA that code for very limited number of proteins, often associated with special functions such as toxins and antibiotic resistance. Plasmids are key molecules for molecular cloning as their size and stability allow for straightforward manipulations and easy propagation.

Gene expression is a two-step process of transcribing the genetic information coded by DNA into a complementary RNA sequence, which is then translated into protein. Studies on DNA bacteriophages were critical for deciphering the underlying principles of the central dogma for the flow of genetic information (DNA→RNA→protein), which later was revised by the discovery of the retroviruses that are able to synthesize DNA using their RNA genomes as templates. Molecular cloning utilizes knowledge and molecular tools related to the flow of genetic information for the purpose of generation of DNA, RNA, and protein molecules with properties of interest. In the discussion to follow the terms molecular cloning, recombinant DNA technology, and genetic engineering will be used interchangeably while referring to manipulation of DNA, although strictly speaking molecular cloning is most frequently used to emphasize techniques, recombinant DNA technology focuses on recombinant DNA molecules produced by molecular cloning, and genetic engineering emphasizes the application of both toward establishing new organismal characteristics with practical value.

2.1.2 Principles of Molecular Cloning

Cloning is the biological process of creating clones of genetically identical organisms, cells, or molecules. Clones of organisms arise via asexual reproduction. For example, each bacterial colony represents a clone comprised of the progeny of a single cell. Spider plants originating from clippings of the same plant are also considered clones. DNA cloning is the process of generating multiple copies of genetically identical DNA molecules, which usually involves insertion of a donor

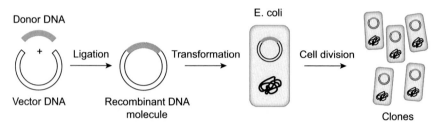

FIGURE 2.2   **DNA cloning overview.** DNA cloning is the process resulting in generation of multiple copies (clones) of genetically identical DNA molecules. Properly processed donor and vector DNA are ligated together and transformed into competent *Escherichia coli* cells. On cellular division, each transformed cell gives rise to genetically identical clones carrying the same recombinant molecule of DNA.

DNA fragment into a vector resulting in a recombinant DNA molecule, which is propagated using *E. coli* (Fig. 2.2). A classical cloning protocol involves (1) digestion of the donor and vector DNA to create compatible ends, (2) ligation, (3) transformation in competent *E. coli* cells, and (4) selection and verification of recombinant DNA molecules. Once the desired clone is selected, needed quantities of the recombinant DNA molecule are obtained by growing larger scale bacterial culture of *E. coli* harboring the molecule of interest. Essentially, the purpose of DNA cloning is to create a DNA clone, which is easy to manipulate and store for future purposes such as sequencing of the donor DNA fragment, mutagenesis, and protein expression. Often the end goal of DNA cloning is the generation of a DNA library, a collection of different DNA clones harboring unique DNA donor fragments cloned in the same vector using identical cloning protocol. The experimental steps are essentially the same as the ones for creating a single clone; however, they are performed simultaneously for multiple donor DNA fragments producing multiple clones. DNA libraries can be created for different purposes such as: (1) to clone and sequence genomes (genome libraries); (2) to clone and identify coding sequences of proteins (cDNA libraries); or (3) to create randomized mutant sets designed to mine for ligands (in phage display) or to select molecules with specific properties (in molecular evolution techniques).

DNA cloning employs many virus-related enzymes to manipulate DNA. Various nonenzyme viral components are utilized as vectors/vector elements, mutagenesis tools, DNA ladders, etc. Currently, three principle approaches for DNA cloning exist: (1) classical cut and paste using restriction enzymes (REs); (2) ligation-independent cloning; and (3) recombination-based cloning. Each one of them will be discussed with emphasis of the relevant viral enzyme(s) and/or components. Sequencing approaches based on viral enzymes are also reviewed.

2.2 VECTORS AND OTHER NONENZYME TOOLS DERIVED FROM VIRUSES

2.2.1 Viral Vectors

Vectors for molecular cloning are small circular molecules of DNA designed to support insertion of donor DNA fragments, proper selection, and independent propagation. Each cloning vector contains a cloning site, composed of DNA sequence harboring multiple unique restriction sites or appropriate sequences compatible with DNA assembly, ligation-independent cloning, or recombination-based cloning. If relevant, signal sequences permitting gene expression from the inserted fragment (promoter, terminator, ribosome binding site, etc.) are positioned in the flanking regions of the cloning site specifically arranged to allow expression in the correct reading frame. Sequences complementary to common sequencing primers are also incorporated in the vicinity of the cloning site, providing straightforward way to verify the sequence of the inserted DNA. The independent propagation of the vector is a function of the available origin(s) of replication. *E. coli* compatible origins of replication ensure that the cloning protocol can take place in *E. coli*. If the vector is to be used in another organism, for example, mammalian or insect cells, it should contain an origin of replication compatible with the host. In many cases, vectors use origins of replication from viruses infecting the host of interest. Selection markers, usually coding for antibiotic resistance, allow for easy identification of cells harboring the vector of interest. It has to be noted that if a vector is to be used in more than one type of organism/cells, selection markers for both *E. coli* and the second host should be present on the vector. In some cases, vectors also contain additional selection markers allowing for easy discrimination between cells harboring empty vector without inserted donor DNA and vector with cloned fragment. These are strategically positioned in the vector, so that the insertion of donor DNA fragment disrupts them enabling visual differentiation such as in blue/white selection when beta-galactosidase (LacZ) is used (for example, pBluescript vector series or the TOPO/Lac Z vector series) or growth-based selection when any of the toxin–antitoxin systems is utilized (for example, the Gateway cloning system). Some vectors are engineered to produce fusion protein of the gene being cloned and a reporter such as Lac Z, green fluorescent protein, or luciferase. Bacterial cells harboring cloned DNA fragments are visualized "biochemically" by the blue color of the product of Lac Z enzyme reaction or by the glowing of the expressed fluorescent proteins.

Several types of vectors are currently in the inventory of molecular cloning tools: plasmids, phages, phagemids, cosmids, and artificial chromosomes. Plasmids were introduced at the beginning of the chapter, and since they are of bacterial origin, they will be discussed only on as needed basis.

For simplicity in the rest of the discussion, the vector DNA sequence is often referred to as a backbone, whereas the donor DNA fragment sequence is referred to as an insert.

Lambda and M13 bacteriophages were among the first vectors developed for molecular cloning and their improvement over time (in comparison with wild type unmodified strains) reflects the advancements of molecular biology knowledge and techniques. Improvements include introduction of unique restriction sites and removal of existing ones; removing nonessential genes to increase insert size capacity, introduction of promoters on both sides of the polycloning site allowing transcription in either direction, development of versions suitable for in vitro transcription and protein expression, development of vectors compatible with mammalian cells, etc. Lambda and M13 bacteriophages can be considered true pioneers in the field of molecular cloning vectors. Bacteriophage lambda's insert size limit is about 20–25 kB, whereas the M13 insert size is about 12 kB. Bacteriophage P1–based vectors can accommodate up to 100-kB inserts (see below). The cloning limits of phage-based vectors are generally determined by the DNA packaging capacity of the relevant capsids and the amount of nonessential genome sequences that can be replaced.

Phagemid vectors (for example, pUC118 and 119) are essentially plasmid vectors containing a phage origin of replication (M13 in the case of the pUC vector series). Their propagation in E. coli is driven by plasmid origin of replication. When cells harboring pUC vectors are infected with a helper M13 phage, the molecules "switch" to phage mode of replication producing ssDNA copies that are packaged into phages. ssDNA isolated from the phages is used for sequencing, site-directed mutagenesis, or as a sequence-specific probe (after labeling) for detecting/analysis of DNA or RNA via Southern and Northern blot, respectively. M13-based phagemid vectors are the vector of choice for generation of phage display peptide/protein libraries. Bacmid vectors are based on the phagemid concept; however, they utilize origins of replication from baculoviruses, which infect insect cells. Bacmids can be propagated both in E. coli and insect cells and are essential tool for generation of recombinant baculoviruses for the purpose of protein expression (discussed in detail in Chapter 4).

Cosmids are vectors designed for cloning of large fragments of DNA and generating genomic DNA libraries. Historically, at the time of their development, they were the vector type with the largest cloning capacity accommodating inserts up to 40–45 kB. Sequence-wise, cosmids can be described as a plasmid containing bacteriophage lambda packaging signals, known as cos sites, thus the name cosmid. Cos sequences are approximately 200 nucleotides long (12 of which are ssDNA protrusions) and are found at the ends of the linear dsDNA genome of the virus. On infection, the cos sites are sealed by the host DNA ligase leading to genome cyclization. Multiple copies of the viral genome are created via rolling circle replication mechanism, which generates long concatemers of head-to-tail oriented genomes. The cos sequences are recognized and cut by the lambda terminase, which introduces single nicks 12 nucleotides apart giving rise to sticky DNA ends. Subsequently, the DNA fragment with resected cos sequence is packaged into lambda capsid. Once recombinant cosmid is generated in a test tube, it can be packaged in a lambda capsid in vitro using packaging lysate of *E. coli* strain harboring lysogenic phage. Cosmids lacking insert are excluded from packaging naturally since their size (4–6 kB) is below the lower limit of a DNA fragments that can be packaged. Cosmids are introduced in *E. coli* via transduction, where they are able to propagate in a plasmid mode driven by plasmid origin of replication. Cosmid-based vectors for mammalian cells have been also developed.

Vectors with even bigger insert capacity are the so-called artificial chromosome vectors, commonly referred to BACs, bacterial artificial chromosomes based on the F-plasmid and accommodating up to 300 kB inserts; PACs, based on bacteriophage P1 replication origins with an insert limit of 300 kB; YACs, yeast artificial chromosomes with cloning limit of 2 MB; and HACs, human artificial chromosomes with a cloning limit up to 10 MB (the size of natural human chromosomes is in the 50–250 MB range). The PAC system takes advantage from the understanding of the life cycle of the bacteriophage P1 and the characteristics of its origins of replication. P1 is a temperate phage and as such it can execute both lytic and lysogenic life cycles. When propagating in lytic mode, the phage uses lytic origin of replication, which directs the synthesis of multiple genome copies connected in a long concatemeric DNA. Single genomes are generated by cuts near the pac (packing) site, and individual genomes are packaged into capsids. When persisting in a lysogenic mode P1 exists as a single copy circular molecule, which replicates only when the host cell replicates. The P1 phage vector system has a cloning limit of 100 kB set by the size of the capsid head and requires elaborate in vitro packaging extracts. In contrast, the PAC system accommodates up to 300 kB and does not pack clones into phages, instead electroporation is used to introduce them into *E. coli*.

Increasing the cloning capacities of vectors for molecular cloning has been essential for enabling science to study large and complex genomes. Other trajectories of vector improvements include the following: (1) increasing/diversifying the functionalities of vectors, for example, including features supporting compatibility with multiple host systems, allowing for protein expression directing polypeptides to different cellular locations, purification, and tracking protein tags, etc.; (2) shortening the time for cloning and reducing the number of individual steps in cloning protocols, for example, the TOPO and Gateway cloning systems; (3) making cloning protocols compatible with high-throughput environments; and (4) eliminating steps that require highly specialized expertise.

Currently, many companies offer preassembled kits with reagents in a user-friendly format that require only the addition of the experiment-specific DNA insert, making cloning appear almost like an exercise of pipetting clear liquids. Molecular cloning truly revolutionized science and continues to push its frontiers. Undoubtedly, viruses, viral components, and host invasion strategies will continue to be part of the process.

2.2.2 Virus-Based DNA Ladders

Studying DNA and developing techniques for nucleic acid manipulations involves many experiments that follow molecule transitions by tracking sizes of DNA fragments in gel-based analyses. DNA ladders are indispensable part of those analyses. Since bacteriophages provide cheap, fast, and straightforward ways to isolate DNA molecules with definite size and characteristics, phage genomes have been used to develop DNA ladders. Fig. 2.3 depicts an example of DNA ladder generated by digestion of lambda genome with a single RE, Hind III, generating distinct fragments with broad range sizes. Restriction digestion of bacteriophage genomes combined with biophysical methods for molecule size determination allowed scientists to develop dsDNA ladders with different ranges before phage genomes were sequenced. In addition, bacteriophages lambda, M13, PhiX174 are routinely used as a source of DNA material/substrates in biochemical, biophysical, and molecular biology research.

2.2.3 Phage-Resistant *Escherichia coli* Strains

Every cloning procedure involves transformation of recombinant DNA molecules in *E. coli* cells to execute clone propagation and selection. Naturally, bacteriophages infecting *E. coli* may present a problem for any effort/process employing the bacterium. The problem can be severe in facilities housing large-scale manufacturing of reagents related to cloning, genomic centers, etc. Standard aseptic molecular biology techniques are an effective tool for control of bacteriophages with exception of bacteriophage T1, which can be spread via aerosols, and it is extremely hard to eradicate as it is resistant to standard autoclaving and bleaching procedures. To avoid the problem with accidental contaminations, most commercially available competent cells are genetically altered to confer T1 resistance by deleting the gene coding for its receptor. T1 binds to ferric heme uptake protein (FhuA, also known as TonA), which happens to be a nonessential gene for *E. coli* and thus can be easily removed. FsuA is also a receptor for bacteriophages T5 and phi80, thus the cells lacking the protein are also resistant to them. Selecting phage-resistant mutants and engineering phage-resistant host cells are critical approaches to maintaining stable starter cultures in dairy and other bacteria-based fermentation industries (Chapter 7).

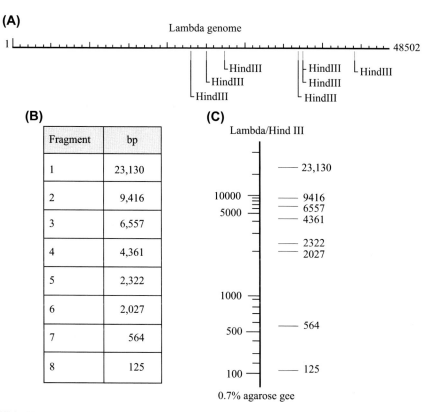

FIGURE 2.3   **DNA ladder derived from bacteriophage lambda.** Bacteriophage lambda has linear dsDNA genome that harbors seven Hind III restriction sides found mostly in the second half of the molecule. Hind III digestion results in 8 distinct fragments ranging between 125 and 23,130 nucleotides, thus generating a broad range DNA ladder. (A) Hind III restriction map of lambda genome; (B) sizes of lamda/Hind III restriction fragments; (C) graphical representation of lambda/Hind III DNA ladder ran on 0.7% agarose gel. The images were generated using bacteriophage lambda sequence and the NEBcutter software (*Vincze T, Posfai J, Roberts RJ. NEB cutter: a program to cleave DNA with restriction enzymes.* Nucleic Acids Res 2003;*31*:3688–91).

2.3 VIRUS-RELATED ENZYMES IN RECOMBINANT DNA TECHNOLOGY

2.3.1 Restriction Enzymes and Viral Nucleases

2.3.1.1 *Restriction Enzymes*

The discovery of REs and ligases, as well as the development of protocols for their manufacturing significantly increased the pace of advancement of molecular biology by allowing controllable and predictable insertion/ deletion of DNA fragments and enabling cut/paste cloning. Although REs are not coded by viruses themselves, their discovery and biological significance are directly connected to virus propagation and virus–host interactions. Biochemically, REs are site-specific endonucleases, which function as a part of the restriction/modification systems in bacteria defending them from invading phages. The first evidence for their existence came from the observation that bacteriophage lambda grows with different efficiency in different *E. coli* strains, some supporting a robust viral propagation, others essentially preventing phage growth. When experimentation showed that phages derived from host strains poorly supporting phage infection have the same growth characteristics as the ones derived from host strains supporting rigorous phage propagation, researchers turned their attention to the host and were able to identify that phage genomes are cleaved in the nonpermissive host strains, suggesting potential involvement of an enzyme with nuclease activity. The discovery drew enormous interest in the research community, which resulted in discovery and characterization of the restriction modification systems in bacteria (Fig. 2.4). Restriction modification systems employ site-specific endonuclease/restrictase and methylase (MTase) enzyme activities with the same specificity. The RE cleaves any unmethylated DNA, i.e., any foreign DNA that enters the cell, by introducing cuts in the phosphodiester backbone of both DNA strands at or near specific target sequence described as a restriction/recognition/target site. The MTase introduces methyl groups at the target sequence into the bacterial genomic DNA, making it inaccessible to the cognate RE. Collectively, the coordinated action of RE/MTase pair neutralizes bacteriophages whose genomes harbor the respective restriction sites while protecting the host genomic DNA from cleavage. Host cells can harbor multiple RE/MTase pairs. As any other host–virus interaction the restriction/modification systems are product of evolutionary tug of war in which both virus and host have evolved different sets of action/counteraction mechanisms. Today, several thousand REs have been described and several hundred of them are commercially available. New England Biolabs Inc. (Ipswich, Massachusetts, USA) maintains an extensive database of REs, REBASE (http://rebase.neb.com), in which new enzymes are added as they are discovered and characterized. Known REs are grouped in four classes based on their characteristics: (1) Type I—cleave at a variable distance from the restriction site; (2) Type II—cleave within or close to the restriction site; (3) Type III—recognize a two-part target site and cleave at specific distance from one of the parts; and (4) Type IV—target only methylated DNA. Molecular cloning techniques utilize predominantly a subgroup of Type II REs, Type IIP, which recognize palindromic target sites. Palindrome sequences are symmetrical and read the same way on both DNA strands. For example, the recognition sequence of EcoRI RE, shown on Fig. 2.5, reads 5′GAATTC3′ on both the top and bottom DNA strands. The names of RE/MTases reflect the host organism they originate from. EcoRI was the first RE isolated from *E. coli*,

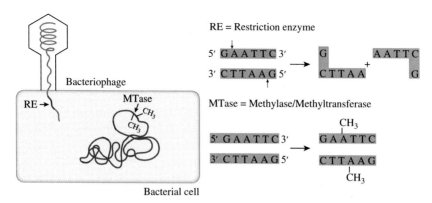

FIGURE 2.4 **Basic concept of bacterial restriction/modification systems.** Restriction/modification systems in bacteria consists of cognate pairs of restriction endonuclease (RE) and methylase (MTase) with identical specificity. Both enzymes recognize the same target sequence cutting the sugar/phosphate backbone (RE) or introducing methyl groups (MTase) at specific residues. REs are capable of destroying the genomes of invading phages containing their target sequence, whereas the host genome methylation prevents the enzyme from digesting the host DNA. Target sites are colored in red. For simplicity only one target site is depicted in methylated state.

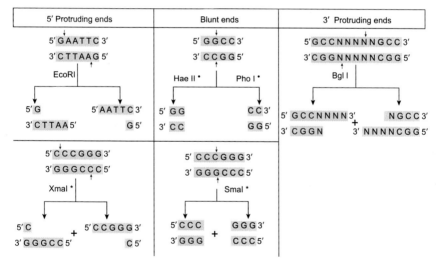

FIGURE 2.5 **Structural characteristics of DNA fragments produced by restriction digest.** Restriction enzymes (REs) are dimeric molecules recognizing target sequences in dsDNA molecules and introducing single breaks in both DNA strands. RE digest products are dsDNA fragments with blunt or sticky/cohesive ends with various lengths 5′ or 3′ ssDNA protrusions (overhangs). The "N" symbol in the recognition site of *Bgl* I RE (top left) designates any nucleotide. Isoschizomers, REs that utilize the same recognition sites, are marked with a dot or an asterisk (*HaeII* and *PhoI* pair or *SmaI* and *XmaI* pair, respectively). Neoisoschizomers, a subset of isoschizomer REs that utilize the same recognition sites and introduce different cut pattern are marked with an asterisk (*SmaI* and *XmaI*). *Arrows* are pointing to the specific location of each RE cut.

strain R, thus its name refers to the genus, species, and the strain name of the bacterium, followed by a number specifying the order of isolation. In contrast EcoRV is the name of the fifth RE isolated from the same host. RE target sites vary widely in length, sequence, and specificity of the introduced cuts. Cuts could be introduced within the sequence resulting in blunt or cohesive/sticky ends with 5′ or 3′ protruding overhangs with different lengths. Some enzymes have identical recognition site and are called isoschizomers, for example, HaeII and PhoI. Isoschizomers that recognize the same target sequence but exhibit different specificity thus resulting in different products are called neoisoschizomers, for example, XmaI and SmaI (Fig. 2.5). The mechanism of REs is extensively studied and efforts to modify their properties have resulted in technological improvements over the years. Type II P REs are working as dimers, which slide along dsDNA until they encounter their recognition sequence. Once bound at the recognition sequence, the enzyme bends DNA, and each monomer introduces a single cut in the phosphodiester backbone, effectively cutting the dsDNA into two pieces. Some REs are not very stringent and can cut at sequences with slight variations from their optimal target site. This activity, known as a "star activity," is often a function of the buffer conditions and the length of the performed digest. Extensive research resulted in engineering of REs with minimal (if any) star activity, and improved overall performance, commonly known as high fidelity or HF REs. Currently, only a subset of the commercial REs are available in HF version; however, their number is increasing. Another improvement is the optimization of the buffer composition and preferences, which allows the use of a single buffer preparation to be used for high efficiency digest of multiple REs, thus facilitating double digest reactions and facilitating automation when the same samples are analyzed with an array of REs. High-purity/high-concentration RE preparations allow for full digest under standard conditions in a matter of minutes, instead of hours. Most REs today are manufactured as recombinant proteins as opposed to being isolated from their original host, thus further facilitating engineering efforts. Attempts have been made to change the specificity of REs in search of approaches to engineer enzymes with new specificities. Successful results from proof of principle experiments have been reported so far only for Type II G enzyme that harbors MTase and endonuclease activities on the same protein chain whose specificity determinants are found in the MTase component. Multiple REs have been engineered as nickases, introducing cut only in one of the DNA strands of dsDNA molecule, thus creating nicked molecules that can be utilized as a starting point for generation of gapped substrates, in studies of strand displacement of polymerases or strand-displacement DNA amplification, studies of exonucleases, optical mapping, among others. Naturally occurring nickase isolated from chlorella virus NYs-1 is also commercially available.

2.3.1.2 *Restriction Enzyme Applications*

The applications of REs are rather numerous (Fig. 2.6), and discoveries made with techniques utilizing them have impact far beyond cloning, genome analysis, and epigenetics. Traditional cloning protocols use cut/paste approach whose success is direct function of the availability of particular RE sites in both vector and donor DNA. Ideally, both molecules are digested with two different unique REs and the generated fragments of interest ligated allowing for straightforward directional cloning, which offers many advantages to downstream applications. Taking into account that there are limited number of vectors and limited number of unique RE sites in each sequence, most cloning procedures require additional steps such as: (1) PCR-based introduction of restriction sites in the donor fragment to ensure compatibility with the vector (2) site-directed silent mutagenesis to introduce or remove restriction sites that may aid or remove obstacles to the cut/paste approach; or (3) intermediate subcloning step(s). REs from Type IIP are preferred REs for cut/paste cloning. Historically, enzymes creating 5′ or 3′ overhang have been used more frequently than the ones generating blunt-end products, due to the higher efficiency of ligation of sticky ends. In some situations, restriction sites convenient for cloning could add 1–2 extra amino acids to either end of the protein coded by the cloned fragment, which may or may not present a complication for downstream applications. For long time, traditional cut/paste cloning with Type IIP REs was the only cloning protocol available; however, newer approaches for specialized purposes have been developed some of which also utilize REs, however, in slightly different context. In vitro assembly approaches to cloning take advantage of the property of Type IIS REs, which cut DNA at a distance from their recognition site. For example, the Golden gate assembly approach utilizes the BsaI RE that recognizes the 5′ GGTCTCNNNNN 3′ sequence (where N can be any nucleotide) and introduces nicks after the

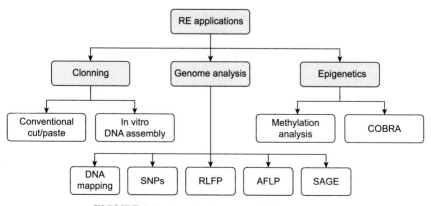

FIGURE 2.6   Applications of restriction enzymes.

first N generating an overhang of four N nucleotides. Multiple fragments with complementary ends (generated by PCR or in vitro synthesis) can be seamlessly assembled into a single larger fragment and subsequently cloned in the vector of interest using the BsaI sites positioned in the flanking sequences. The approach is used to simultaneously introduce multiple site-directed mutations in different parts of the same protein, to design sequence-specific TALENs (transcription activator-like effector nucleases) for genome editing (Chapter 3), among others. The technique was used to assemble the mouse mitochondrial genome (16.3 kB) in vitro from 60-mers, as well as a step in the assembly of the genome of *Mycoplasma mycoides* (1.1 MB), which was transplanted to an *Mycoplasma capricolum* to yield new self-replicating *M. mycoides* cells. The assembly cloning approaches use vector cut with RE, which is ligated to large DNA fragment generated via assembly of overlapping shorter fragments.

In addition to cloning, REs are indispensable reagent for genome analysis even in the age of fast and relatively low-cost sequencing. The first physical genome maps were records of genomic DNA digested with single RE, similar to the map of lambda genome shown in Fig. 2.3. Not surprisingly, viral genome maps were among the first generated simply because virus preparations offered feasible material to isolate large amounts of intact genomic DNA suitable for such analysis. RE DNA mapping is based on RE digest of genomic DNA followed by electrophoretic analysis of the resulting fragment. Fragment mobility was used to generate a linear map of the genome and allowed for future organization of genetic markers, sequences, etc. Bacteriophage genome and plasmid maps accounting for the position of various restriction sites were essential for the progress of molecular cloning technology before sequencing was easily accessible. Today, such maps are a useful planning tool accompanying any cloning manual and easily generated by various software programs. DNA fingerprinting, one of the very few molecular techniques general public is well aware of, is conceptually based on the principle of DNA mapping. Single nucleotide polymorphism (SNP) is a powerful approach to genotyping, which employs PCR to amplify short DNA fragments containing loci with known alternative sequences. The resulting fragments are digested with RE, and the outcome of the reaction is monitored by electrophoresis. The RE-base version of SNP has limited utility, simply because not every SNP is associated with a restriction site. Restriction fragment length polymorphism (RFLP) analysis combines the concepts of DNA mapping and Southern blot. Genomic DNA is digested with a RE, and the resulting products are resolved by electrophoresis and transferred to a membrane. A labeled probe containing the sequence under investigation is prepared and hybridized to the membrane generating a specific band pattern. In addition to DNA fingerprinting, the method is used to compare cultivars and animals in agricultural breeding programs, as well as for purity control in seed production. The classic RFLP approach requires large amount of genomic DNA and tends to be cumbersome. More sensitive alternative is amplification fragment length polymorphism (AFLP) in which REs generated genomic DNA fragments are ligated to adapter sequences with complementary overhangs and then amplified by PCR with primers annealing to the adapters. The PCR products are resolved on gel and visualized by autoradiography or fluorescence. AFLP fragments can be easily subjected to sequencing if necessary. The method requires only nanogram amount of genomic DNA and is used in plant and animal breeding programs, molecular phylogeny, as well as to generate maps of organisms whose genome has not been sequenced. AFLP detection can be automated if the protocol is executed with fluorescently labeled primers and capillary electrophoresis is used to resolve the PCR-amplified fragments.

Serial analysis of gene expression (SAGE) technique probes gene expression on genomic level by creating series of linked tags (Fig. 2.7). The method is based on the assumption that each tag can reliably identify the originating transcript once tag sequences are determined. Conceptually, SAGE answers the same type of experimental question as microarrays, however, does not require unique equipment (DNA chip and relevant reader) and relies on sequencing rather than probe hybridization. SAGE is often used to compare the expression profiles of different states of the same system such as healthy versus infected cell, healthy versus cancerous cell, etc. Polyadenylated mRNAs are extracted from sample(s) of interest and converted to cDNA using biotinylated oligoT primer, which anneals to the polyA tail of all mRNAs. The cDNA molecules are attached to streptavidin beads and digested with a RE of choice (designated as anchoring enzyme [AE]) recognizing four-nucleotide long restriction site. Since such REs cut quite frequently every single cDNA would be cut at least once. The digestion results in streptavidin/biotin immobilized cDNA fragments with variable length corresponding to the 3′ end of the original mRNAs, all ending with identical overhang generated by the anchoring RE. At the next step the sample is split into two, and each pool is ligated to an adapter (adapter A or adapter B) containing an additional site for an RE, designated as a tagging enzyme (TE), which cuts at a distance from its recognition site. Digestion with the TE releases the cDNA fragments from the streptavidin beads in solution. The overhangs created by the TE digestion are blunted and cDNA tags from pool 1 ligated to cDNA tags from pool 2 resulting in ditag fragments in which a random pair of two cDNA tags are flanked by AE/TE restriction sites and adapter A or B, respectively. Primers complementary to the adapter sequences are used to amplify the ditags in a bulk PCR reaction. The PCR products are then digested with the AE producing ditags with sticky ends and releasing the adapter pieces. Next, the ditags are ligated generating a concatemer of tags, which is sequenced and sequencing results are analyzed computationally to

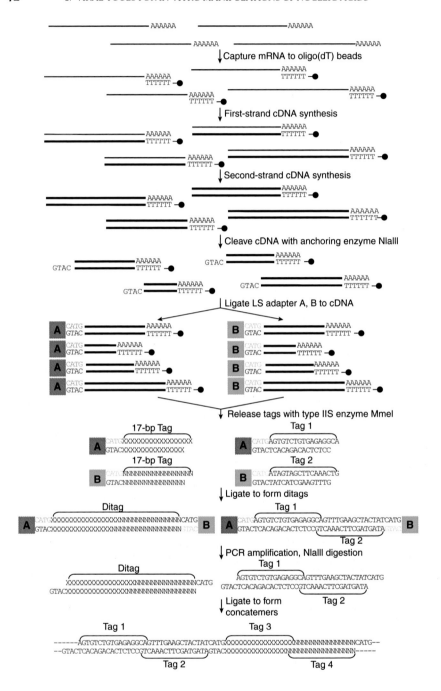

create an expression profile. mRNAs with higher level of expression are found to have proportional tag representation in the sequenced pool of ditags. Major advantage of the technique is its speed, sensitivity, and the bulk processing of the samples, which allows quick deciphering of expression profiles with applications in diagnostics of complex diseases such as cancer. The technique does not account for cDNA without AE restriction sites and does not distinguish between transcripts that generate the same tag sequences. Variations of SAGE such as longSAGE, superSAGE, and deepSAGE explore different REs to create longer tags and resolve some of the limitations of the original SAGE protocol. The concept of using specific RE and an adapter with corresponding target site is employed in multiple techniques where enrichment and/or amplification of the DNA samples is beneficial, such as some of the next-generation sequencing technologies, libraries for genotyping by sequencing, identification of cross-linking sites of proteins and DNA in chromosome mapping, etc.

REs are key reagents in analyses of DNA methylation in epigenetics research and laboratory techniques employing REs ability to differentiate between different methylation states (for example, QuickChange mutagenesis technology). DNA methylation analysis combines DNA mapping concepts with methods to specifically identify base methylation. Essentially, restriction digests with methylation-sensitive and insensitive isochisomers are performed, and the resulting restriction patterns are compared using gel electrophoresis, Southern blot or DNA amplification–based techniques. The higher the differences between both restriction profiles are, the higher the level of methylation is. Methylation is associated with repression of transcription and is directly connected to programmed changes in gene expression during development and X-chromosome inactivation, as well as with altered gene expression in cancerogenesis. Only two of the four DNA bases are subject to methylation in vivo: cytosine and adenine; however, cytosine methylation is more widely spread across the living world and main focus of attention. The degree of genome methylation varies widely between species. For example, only 0.3% of *Drosophila* genome was found to be methylated, whereas the methylation

FIGURE 2.7 **Overview of SAGE (serial analysis of gene expression) technique.** SAGE is a technique allowing *en mass* analysis of mRNA expressed in a sample of interest. The protocol includes the following steps: (1) capture mRNA with oligo(dT) magnetic beads; (2) first-strand cDNA synthesis; (3) second-strand cDNA synthesis; (4) cleave cDNA with anchoring enzyme *NlaIII*; (5) split samples and ligate to LS adapter A or B with *NlaIII* overhang (CATG); (6) combine samples and cleave with tagging enzyme MmeI to release 17-bp tags, examples of which are shown (tags 1 and 2); (7) ligate to form ditags; (8) PCR amplification and NlaIII digestion, releasing ditags from adapters; and (9) ligate ditags to form concatemers, which can be cloned and sequenced. *Image and figure legend reprinted from Hu M, Polyak K. Serial analysis of gene expression. Nat Protoc 2006;**1**(4):1743–60, with permission from Nature publishing group.*

of *Arabidopsis thaliana* genome could be as much as 14%. In mammals, cytosine methylation almost exclusively takes place in the context of CpG (5'C-phosphate-G3') dinucleotide. The designation CpG was introduced to specify a dinucleotide of C and G connected to each other with a phosphodiester bond, as opposed to GC base pair where both complementary bases are interacting with each other with hydrogen bonds and are not connected covalently. Cytosine is methylated in position 5, i.e., at the same atom as thymine, on both strands of the CpG sequence.

One approach to methylation analysis is the so-called HELP assay (HpaII fragment Enrichment by Ligation-mediated PCR assay), which takes advantage of the isoschizomer pair HpaII and MspI. Both enzymes recognize the 5'-CCGG-3' target sequence; however, HpaII cuts it only if the CpG dinucleotide in the middle is unmethylated, whereas MspI cuts both methylated and unmethylated tetranucleotide. The digested DNA is ligated to adapter and subsequently amplified with PCR. The fragment polymorphism in both digestion reactions is compared and methylation quantified. The degree of digest in the HpaII sample is used as a measure of methylation, whereas the degree of digest in the MspI reaction is used as a reference point for copy number and control for the integrity of the process. Sequencing of the PCR fragments can provide further information for the location of the methylation modifications. The assay can be applied to regions with different lengths ranging from a single locus to the entire genome; however, as the complexity of the sample increases the interpretation of the data becomes more complex. As a result the HELP assay is rarely applied to whole genomes.

Alternative technique for methylation analysis is COBRA (combined bisulfite restriction analysis), which takes advantage of the bisulfite deamination of cytosine. Treatment of DNA with sodium bisulfite results in conversion of cytosine to uracil, whereas methylcytosine remains unchanged. The treated DNA is then subjected to PCR with primers lacking CpG dinucleotides and the resulting PCR products digested with REs and restriction fragments analyzed and quantified. TaqI (TCGA) and BstUI (CGCG) enzymes are usually used in COBRA assay, which would cut any sequence that was methylated and leave the unmethylated sequences intact since they were converted to TTGA and TGTG as a result of the bisulfite treatment (C to U chemical conversion) and PCR (U will be replaced with T since the PCR reaction contains dTTP and not dUTP). A digestion with Hsp92II (CATG) is carried out in parallel as a reference for the level of bisulfite conversion. Major advantage of the method is that it uses very small amount of DNA (the method was originally developed for methylation analysis in paraffin-embedded tissue slices). The major disadvantage of the method is its reliance of specific REs thus detecting methylation status only of CpG dinucleotides that are part of the recognition sequence of the RE of choice. In most cases,

COBRA is used as a quick assay to probe for changes in methylation and alternative methods are used for complete evaluation. The method is quantitative, inexpensive, and scalable, thus ideal for high-throughput format of sample analysis.

The DpnI RE, which digests the 5′-GATC-3′ sequence only when methylated or hemimethylated (adenine methylation), is used in several site-directed mutagenesis approaches (for example, QuikChange and Q5 site-directed mutagenesis kits) as a tool to distinguish between parental/template nonmutated DNA and newly synthesized mutated DNA. Mutations (substitution, deletions, and insertions) are designed as part of the primer used to copy the template by high fidelity DNA polymerase. The protocol uses multicycle isothermal amplification that generates a mix of nonmethylated product carrying the mutation(s) of interest, hemimethylated product containing one template strand and one newly synthesized strand and fully methylated molecules of unused template. DpnI treatment digests the methylated and hemimethylated molecules and does not impact the nonmethylated ones. Subsequent transformation allows them to be propagated and clones carrying the mutation(s) of interest to be selected.

A unique RE-based approach to mutagenesis beyond methylation is the restriction enzyme–mediated integration used for genetic studies in fungi and other organisms. In particular, it was used to map development pathways in *Dictyostelium* and to dissect virulence genes in fungal plant pathogens. Its mechanism is somewhat reminiscent of transposon insertional mutagenesis. A plasmid linearized with RE is introduced in the cell of interest together with the RE used for linearization, which cuts the targeted genome at random position containing the relevant recognition site. The linear plasmid is then inserted presumably by the nonhomologous end-joining pathway of the cell. If the insertion takes place within a coding sequence, the gene is disrupted, thus enabling researches to gain knowledge of its function. REs are useful tools for molecular cloning and molecular biology in general, offering fast, easy, and straightforward way to manipulate and analyze DNA molecules.

2.3.1.3 Sequence-Nonspecific DNA Nucleases

Many viruses code for sequence-nonspecific nucleases with various properties that are essential for viral propagation and function in processes such as DNA replication, repair, and recombination. For example, bacteriophage T7 codes for a pair of exonuclease and endonuclease, gp6 and gp3, postulated to degrade the host genome in order its building blocks to be used to synthesize phage genomes. Bacteriophage lambda, herpes viruses, baculoviruses code for nucleases that promote efficient recombination in tandem with ssDNA-binding proteins. Recombineering using the lambda Red system and other in vivo approaches involving nucleases are described in Chapter 3. In vitro applications of viral nucleases depend on their specificity and their uniqueness as many cellular nucleases have been described and characterized before or in parallel to the discovery of viral nucleases. Viral nucleases are used predominantly as a tool to generate various DNA molecules for the purpose of biochemical analyses or to characterize the chemical nature of products of reactions involving DNA manipulations. For example, the bacteriophage T5 D15 gene codes for $5'-3'$ exonuclease, commonly referred to as T5 exonuclease, which degrades linear ss and dsDNA molecules and nicked plasmids, but cannot utilize supercoiled plasmids as a substrate. T5 exonuclease is an excellent choice for preparing supercoiled DNA. T5 exonuclease is also employed in the Gibson assembly cloning approach. In brief, long DNA fragments are amplified as overlapping fragments. The primers annealing to the far left and the far right of the DNA template contain short sequences complementary to the ends of the cloning vector linearized with RE. The T5 exonuclease partially digests the ends of the vector and inserts molecules creating ssDNA overhangs that facilitate sequential assembly of the vector backbone and the multiple DNA inserts resulting in one circular molecule with gaps. The polymerase and ligase in the reaction mix fill the gaps and seal the fragments, thus producing seamless DNA molecule. Bacteriophage T7 $5'-3'$ exonuclease, coded by gene 6, has no affinity for ssDNA, and it is mainly used to degrade dsDNA nonspecifically. It can also digest RNA or DNA from RNA/DNA hybrids with low processivity.

Bacteriophage T7 endonuclease I, coded by gene 3, is used for generation of random fragments of linear dsDNA for the purpose of shotgun cloning. It also recognizes and cleaves cruciform DNA, Holiday junctions and structures, mismatched and heteroduplex DNA, and it is used for detection of the above structures, resolution of branched DNA and cleavage of heteroduplex DNA. The later make it a useful reagent for analysis of recombination and replication intermediates. The enzyme is also used to detect mutations, and the importance of this application is on the rise in the context of the current advancements of in vivo gene-editing technologies such as CRISPR/Cas, TALENs, and zinc finger nucleases. Genomic DNA from cells that were subject to editing is isolated and the region of interest is amplified by PCR. The goal is to generate a PCR product with a length around 1 kB harboring the site of potential mutation in a position that will allow easy differentiation between uncut and cut DNA by the size of the cleavage-generated products. PCR DNA fragments are melted and reannealed potentially generating three types of products: (1) wild-type DNA duplex, (2) mutant DNA duplex, and (3) heteroduplex of wild-type and mutant DNA strands annealed together. Treatment with T7 endonuclease I results in cleavage of the heteroduplex, which is visualized by electrophoresis. Quantification of the amount of each product can be used for calculation of mutation frequencies. Bacteriophage T4 endonuclease

VII, coded by gene 49 and also known as T4 resolvase, has similar properties to T7 endonuclease I and can also be used to detect mutations. T4 endonuclease V, also known as pyrimidine dimer glycosylase, recognizes pyrimidine dimers and removes them by the combined action of its glycosylase and endonuclease activities. The enzyme is used in DNA damage studies. Although the applications of sequence nonspecific nucleases are not as extensive as those of REs, they have their place in vitro manipulations of DNA.

2.3.2 Replication-Related Enzymes

2.3.2.1 DNA Polymerases

DNA polymerases (DNA pols) are diverse group of enzymes functioning in DNA replication, recombination, and repair. Currently known DNA pols are classified in seven different families based on their phylogenic relationships to *E. coli* and human DNA pols (including one family of reverse transcriptases). Replicative DNA pols exhibit two major activities: polymerase and $3'-5'$ exonuclease/proofreading (Fig. 2.8). In addition, some enzymes have $5'-3'$ exonuclease and strand displacement activities.

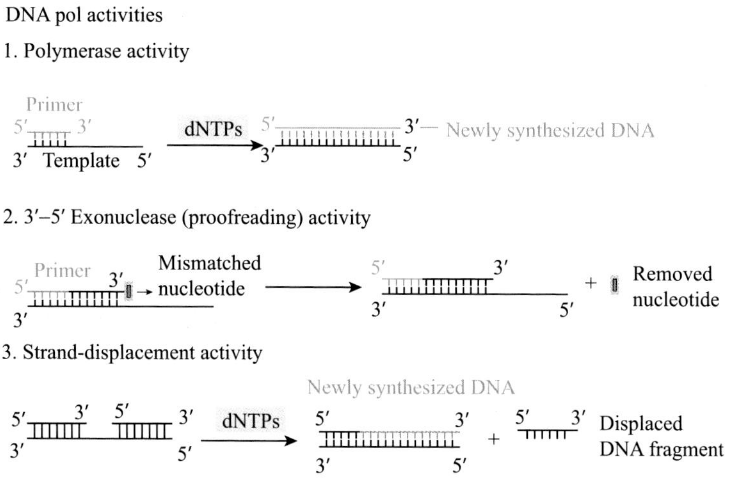

FIGURE 2.8 **Biochemical activities of DNA-dependent DNA polymerases.** DNA-dependent DNA polymerases catalyze polymerization of dNTPs into polynucleotide chain by adding one nucleotide at the time to the free $3'$ hydroxyl group of the last nucleotide while following the rules of complementarity (1). If a nucleotide is added erroneously, it is removed by the $3'-5'$ exonuclease activity of the enzyme (2). Some DNA polymerases exhibit strand displacement activity allowing them to displace annealed DNA fragments and replace them with a continuous DNA strand (3). DNA polymerases are unable to initiate de novo synthesis and require a primer (depicted in blue) providing a free hydroxyl group to allow DNA synthesis initiation.

Strand displacement activity allows polymerases to continue adding nucleotides when they encounter a dsDNA area of the template. Instead of stopping and dissociating the molecule displaces the DNA strand annealed to the template without interrupting polymerization. Fig. 2.8 shows an example of short DNA fragment being displaced; however, enzymes with a robust strand-displacement activity can display kilobases of DNA when working around a circular template. DNA pols cannot initiate de novo synthesis and all utilize primers providing a 3'-OH group to start the synthesis. Both RNA and DNA primers are utilized and some viruses, for example, phi29, use a protein primer. DNA polymerization is characterized by its speed, fidelity, error rate, and processivity. Fidelity is the ability of the enzyme to replicate DNA template precisely. HF polymerases have low error rates (number of misincorporated nucleotides per unit length) and efficient proofreading. Processivity is defined as the ability of the enzyme to continuously polymerize without releasing the template strand. Both HF and processivity are desired properties of DNA pols used as molecular tools since they offer the most possibility for stringent control. DNA pol properties can be modulated by optimizing buffer conditions (especially the concentration of metal ions), genetic engineering, and molecular evolution.

The best characterized phage DNA pols belong to the A family (*E. coli* Pol I like), such as T3, T5, T7 DNA pols, or to the B family (*E. coli* Pol II like), such as T4, phi29. Unlike cellular DNA pols, phage replicases are monomeric or have limited number of subunits and employ small number of accessory factors (if any). Polymerases of many eukaryotic viruses have been studied and are considered targets for relevant drug development; however, their applications as molecular biology tools are limited to the studies of the virus they originate from.

The most significant applications of phage DNA pols are related to DNA sequencing and rapid DNA amplification (T7 DNA pol/Sequenase and phi29 DNA pol). In addition, T7 and T4 DNA pols are used to generate DNA substrates for biochemical studies or for preparation of labeled DNA probes. T4 DNA pol is also used in ligation-independent cloning protocols (Fig. 2.9). The first step in the technique is based on the 3'→5' exonuclease activity of the enzyme. Vector molecules linearized with REs and PCR products with overlapping ends are treated with T4 DNA pol in the presence of a single dNTP in the reaction mix. Under the conditions of the reaction the T4 DNA pol exonuclease activity hydrolyzes nucleotides creating cohesive ends. Next, vector and insert molecules are annealed creating circular dsDNA molecules with gaps in the vector/insert junctions. Finally, a full set of dNTPs is provided and the polymerase function of the enzyme fills in the gaps. The generated nicked molecules are transformed in *E. coli* and the nicks sealed in vivo.

Applications of viral DNA pols in sequencing have evolved dramatically head to head with advancements in the fields of nucleotide substrates

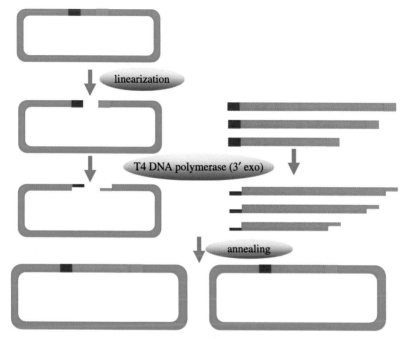

FIGURE 2.9 **Applications of T4 DNA polymerase (DNA pol) in ligation-independent cloning.** Linearized blunt-end vectors and PCR-amplified target gene containing complementary tails are digested by the $3'{\to}5'$ exonuclease activity of T4 DNA pol exonuclease activity in the absence of full set of dNTPs. Vector and fragment are annealed and gaps are filled in by the polymerase activity of T4 DNA pol resulting in dsDNA circular molecule with nicks at the vector/insert junctions. Nicks are repaired in vivo after transformation in *Escherichia coli*. *Image courtesy Jia B, Jeon CO. High-throughput recombinant protein expression in* Escherichia coli: *current status and future perspectives.* Open Biol *2016;6(8):160196.* https://doi.org/10.1098/rsob.160196, *CCBY,* http://creativecommons.org/licenses/by/4.0/.

and equipment technologies supporting the process. Initially, enzyme-driven sequencing employed chain termination as conceptual principle to obtain DNA sequences, which was later replaced by sequencing by synthesis concept and unconventional uses of phi29 DNA pol in nanopore sequencing as enzyme-threading DNA. Sequencing by chain termination (Fig. 2.10), also known as Sanger method was developed with the Klenow fragment of *E. coli* DNA pol I, a proteolytic fragment with polymerase and proofreading activity but not $5'{\to}3'$ exonuclease activity. Sequences were derived from a set of four individual reactions each containing the same template, the same radioactively labeled primer, mix of four dNTPs, and one ddNTP. After completion the four reactions were analyzed on gel and sequences deciphered based on the gel mobility of the generated chain-terminated DNA fragments. The reads were relatively short and with low quality due to low enzyme processivity,

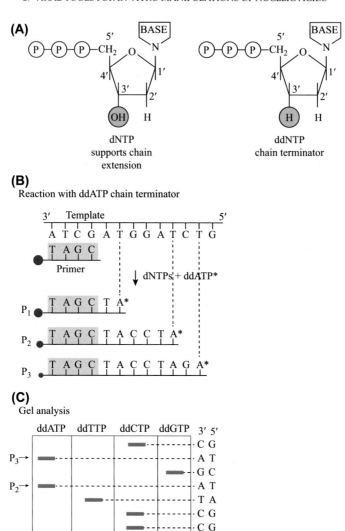

FIGURE 2.10 **Conceptual principle of sequencing by chain termination.** (A) Chemical structures of natural (dNTP) and chain terminator (ddNTP) nucleotides. The critical difference in position 2′ is highlighted in orange. (B) Example reaction with ddATP chain terminator. The *black circle* depicts the end label of the utilized primer (highlighted in gray). The asterisk emphasizes the incorporation of a chain terminator. (C) Gel analysis of four sequencing reactions of the same template, each containing a single-chain terminator. Fragments labeled as P1, P2, and P3 correspond to the P1, P2, and P3 products depicted in panel b. The shorter the product is, the faster it migrates thus bands at the bottom of the gel correspond to the sequences polymerized at the beginning of the assay (3′ end of the template/5′ end of the product). Gel electrophoresis is supplemented by autoradiography to visualize the products of each sequencing reaction.

high level of discrimination between dNTPs and ddNTPs (ddNTPs were added up to 1000 times slower), and some degree of bias toward specific ddNTPs, resulting in bands with uneven intensity. In part, both of these problems were solved by replacement of the Klenow fragment with bacteriophage T7 DNA pol, coded by T7 gene 5. T7 DNA pol is rather interesting enzyme. It incorporates ddNTPs fairly fast, only a couple of times slower that dNTPs, thus allowing lower concentration of ddNTPs to be used in the sequencing reaction effectively lowering cost and increasing quality of sequencing. In addition, when in complex with its processivity factor, *E. coli* thioredoxin, T7 DNA pol has fairly high processivity allowing for longer reads. Buffer optimization studies led to selection of buffer conditions (Mn^{2+} containing isocitrate buffer) that further minimized the differences in rate of incorporation between regular and chain-terminating substrates. Further improvements involved chemical modification of the enzyme molecule to reduce the proofreading activity of the enzyme, which impacts negatively the length of the sequencing reads (as the concentration of free dNTPs decreases the differences between the rates of polymerization and degradation balance out resulting in no net change of the length of the synthesized polynucleotide). The improved version of T7 Pol, product of the work of Stanley Tabor and Charles Richardson (Harvard Medical School), became the first available straightforward to use sequencing enzyme under the commercial name Sequenase (USB Corporation). Further research efforts resulted in the identification of exonuclease deficient mutants, which were the basis for Sequenase 2.0. Quality of sequencing was further improved by the addition of pyrophosphatase in the reaction mix. The latter breaks down the pyrophosphates released during polymerization and influences the equilibrium between polymerization and pyrophosphorylysis, thus favoring nucleotide addition and longer reads. Studies of the mechanism of discrimination between dNTPs and ddNTPs among different polymerases identified that T7 phenylalanine at position 526 was responsible for the phenomenon and that F526Y mutation abolishes the difference in rates of incorporation between both types of nucleotides. Based on sequence homology the F526Y mutation was "transferred" into Taq polymerase (F667Y) drastically changing its properties. The modified Taq polymerase had two fold preference for ddNTP compared with dNTP completely reversing the natural several 1000-fold preference for dNTP. Remarkably, all of the above discoveries were made without any structural information about T7 DNA pol. In addition, to the improvements of the DNA pol, sequencing by chain termination benefited from the introduction of fluorescently labeled primers corresponding to each of the A, G, T, and C reactions and the development of computer-driven fluorescence detection system. Collectively, both advancements allowed for single reads of the mix of four reactions (each nucleotide was specified by the color corresponding to the label of the primer utilized in the reaction). At the next step of technological improvement, fluorescently labeled ddNTPs and capillary electrophoresis came to play, which allowed full automatization of the sequence reading (Applied Biosystems/Life Technologies genetic analyzer technology). Although Sanger sequencing produces long reads of high quality, the technical aspects of capillary electrophoresis cannot deliver the capacity, speed, and cost matching the needs of whole genome sequencing. Next-generation sequencing technologies (for example, Illumina, Qiagen/IBS) employ reversible fluorescent dye terminators and stepwise sequencing by synthesis principle. In stepwise sequencing by synthesis protocol (Fig. 2.11) sequencing is accomplished by continuous cycling through the following steps: (1) addition of new reversible chain terminator nucleotide (all NTPs are added in a single reaction, each labeled with different color fluorescent dye), (2) imaging of the extended product, and (3) chemical removal of the chain-terminating features restoring the 3′OH and allowing the addition of the next

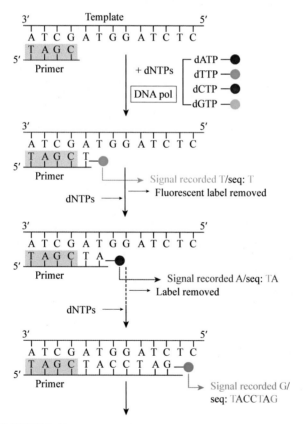

FIGURE 2.11   Conceptual principle of sequencing by synthesis.

nucleotide. These advancements allowed sequencing to gravitate away from capillary electrophoresis, thus making possible millions of different DNA molecules to be sequenced simultaneously. Archeal DNA pols (not viral) were found to be suitable for incorporation of the new-generation reversible chain terminators. The chemistries of unblocking the chain terminators leave leftover chemical groups in the synthesized product, which may interfere with the sequencing process. Fluorescently labeled polyphosphate nucleotides were developed in attempt to achieve seamless synthesis. These nucleotides are quite bulky and were not a good substrate for any of the previously employed DNA pols; however, the bacteriophage phi29 DNA pol readily utilizes them. Phi29 DNA pol has great features for sequencing enzyme: it is monomeric and polymerizes DNA with high fidelity and extreme processivity. Collectively, the fluorescently labeled polyphosphate nucleotides and phi29 DNA pol allowed for the development of newer generation of sequencing technologies based on real-time single-molecule sequencing by synthesis (Fig. 2.12). In the SMRT (single-molecule real-time sequencing, Pacific Biosciences), single DNA pol molecules are immobilized on a hard surface in miniature chambers together with template DNA. The binding of each fluorescent polyphosphate nucleotide is imaged before the phosphodiester bond is formed and the fluorescent label released. Captured fluorescent signal is processed computationally to decipher the order of nucleotides.

Nanopore-based sequencing (Oxford Nanopore technologies) employs the phi29 DNA pol in slightly different capacity. Biological nanopores (Alpha-hemolysin or MspA proteins) are inserted in membrane layer separating two chambers with electrolyte solution. A single molecule of phi29 DNA pol is tethered to the pore guiding the newly synthesized DNA through the channel of the molecule. The chemical nature of the passing nucleotide(s) is determined based on the changes in the electrical current. Electrical signals corresponding to single nucleotides or various k-mers are correlated to their sequences and used as a standard to decipher the DNA sequence from the ONT output (Fig. 2.12B). Nanopore sequencing allows for automated sequencing in parallel and for long reads (sequencing of the lambda genome, approximately 50 kB, in one run has been reported), which are very advantageous characteristics in the current state of the sequencing field.

The high processivity and the strand displacement properties of phi29 DNA pol are utilized in rolling circle amplification technique, which is used for the preparation of DNA nanoballs used in ligation-mediated sequencing (described in Section 2.3). The technique is also used for detection of whole viral genomes in diagnostics, for phylogenetic analyses, epidemiological studies and studies of genome organization and is generally considered as an isothermal alternative to PCR amplification.

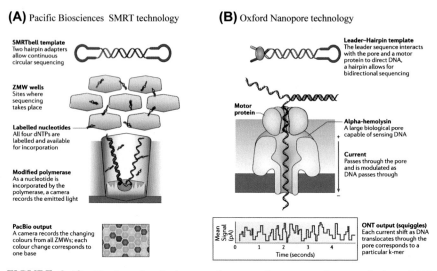

FIGURE 2.12 **Single-molecule long-read sequencing approaches employing phi29 DNA polymerase (DNA pol).** (A) Single-molecule real-time (SMRT) sequencing from Pacific Biosciences (PacBio). Template fragments are processed and ligated to hairpin adapters at each end, resulting in a circular DNA molecule with constant single-stranded DNA (ssDNA) regions at each end with the double-stranded DNA (dsDNA) template in the middle. The resulting "SMRTbell" template undergoes a size-selection protocol in which fragments that are too large or too small are removed to ensure efficient sequencing. Primers and an efficient phi29 DNA pol are attached to the ssDNA regions of the SMRTbell. The prepared library is then added to the zero-mode waveguide (ZMW) SMRT cell, where sequencing can take place. To visualize sequencing, a mixture of labeled nucleotides is added; as the polymerase-bound DNA library sits in one of the wells in the SMRT cell, the polymerase incorporates a fluorophore-labeled nucleotide into an elongating DNA strand. During incorporation, the nucleotide momentarily pauses through the activity of the polymerase at the bottom of the ZMW, which is being monitored by a camera. (B) Oxford Nanopore Technologies. DNA is initially fragmented to 8–10kB. Two different adapters, a leader and a hairpin, are ligated to either end of the fragmented dsDNA. Currently, there is no method to direct the adapters to a particular end of the DNA molecule, so there are three possible library conformations: leader–leader, leader–hairpin, and hairpin–hairpin. The leader adapter is a double-stranded adapter containing a sequence required to direct the DNA into the pore and a tether sequence to help direct the DNA to the membrane surface. Without this leader adapter, there is minimal interaction of the DNA with the pore, which prevents any hairpin–hairpin fragments from being sequenced. The ideal library conformation is the leader–hairpin. In this conformation the leader sequence directs the DNA fragment to the pore with current passing through. As the DNA translocates through the pore, a characteristic shift in voltage through the pore is observed. Various parameters, including the magnitude and duration of the shift, are recorded and can be interpreted as a particular k-mer sequence. As the next base passes into the pore, a new k-mer modulates the voltage and is identified. At the hairpin, the DNA continues to be translocated through the pore adapter and onto the complement strand. This allows the forward and reverse strands to be used to create a consensus sequence called a "2D" read. *Figure and figure legend reprinted from Goodwin S, McPherson JD, McCombie WR. Coming of age: ten years of next-generation sequencing technologies. Nat Rev Genet 2016;**17**(6):333–51, with permission from Nature publishing group.*

2.3.2.2 *Reverse Transcriptases*

Reverse transcriptases (RTs) are RNA dependent DNA pols initially iso-lated from retroviruses. In addition, RTs are coded by dsRNA viruses that utilize reverse transcription such as hepatitis B virus (replication of hepati-tis is discussed in Chapter 1); and various retroelements in eukaryotes and prokaryotes. The enzyme telomerase maintaining the ends of the eukary-otic chromosomes is technically also a reverse transcriptase, although its mechanism is very distinct from conventional RTs. Historically, the dis-covery of RT revolutionized molecular biology leading to the revision of the central dogma and enabling scientists to develop new research tools that heavily influenced cloning, analysis of gene expression and RNA biology. HIV RT is one of the most extensively studied polymerases in the context of understanding the biology of this devastating virus and design-ing RT inhibitors as drugs to manage HIV infections. RTs exhibit three key enzyme activities (Fig. 2.13): (1) RNA-dependent DNA pol that uses ssRNA template and a primer (tRNA^{Lys} for HIV RT) to synthesize ssDNA/cDNA, which remains hybridized to its RNA template; (2) RNAse H endonucle-ase, which selectively degrades the RNA strand of DNA/RNA hybrids and (3) DNA dependent DNA polymerase activity, converting the single-stranded cDNA into dsDNA. Conventional RT enzymes have two active sites, one executing the polymerase activities and another executing the endonuclease activity. RT are monomeric or dimeric proteins and some lack intrinsic RNAse H activity. The RTs of Moloney murine leukemia virus (M-MLV) and Avian myeloblastosis virus (AMV) are most frequently used as molecular tools in RT-PCR, RT-qPCR, cDNA cloning, RNA sequencing and any other experimental technique/approach that requires conversion of RNA to DNA. Site-directed mutagenesis and protein evolution have been utilized to optimize those enzymes improving thermostability and

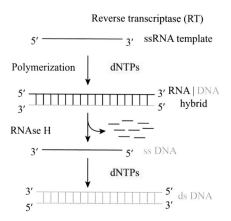

FIGURE 2.13   Reverse transcriptase activities and mechanism of action.

modulating RNAseH activity. Using thermostable version of RT is beneficial for lowering the nonspecific nucleic acid amplification and minimizing impact of complex secondary structures. Robust RNAseH activity is an advantage in RT-PCR, whereas lower RNAseH activity is beneficial in cDNA cloning protocols, especially when very long mRNA transcripts are reverse transcribed. In some cases RT is used only to produce the RNA/DNA hybrid and a conventional DNA pol carries out the cDNA to dsDNA polymerization step.

2.3.2.3 DNA Ligases

DNA ligases are best known for their role in joining adjacent Okazaki fragments at the lagging strand of the replication fork; however, they are essentially involved in any process that requires sealing of phosphodiester bonds from the DNA backbone. Ligases catalyze the formation of phosphodiester bond between 3'OH and 5'phosphate on various substrates such as DNA nicks, DNA fragments with various lengths cohesive ends, DNA fragments with blunt ends and some DNA/RNA hybrids. Fig. 2.14 depicts the basics of ligase action at a DNA nicked substrate. The most widely used viral ligase is the T4 DNA ligase, which is routinely employed in cloning applications. T4 DNA ligase exhibits highest activity ligating DNA nicks and fragments with overhangs longer than 2 nucleotides, whereas fragments with blunt ends, shorter overhangs or nicks containing mismatches are processed less efficiently requiring higher enzyme concentrations and extended ligation times. Interestingly, T4 ligase can also join the 3' end of RNA strand to a DNA strand with 5' phosphate when the DNA fragment is annealed to a complementary DNA. Many commercial formulations of T4 ligase are available, which differ in enzyme concentration, presence of proprietary ligase activity enhancers, or buffer conditions compatible with various downstream applications such as transformation of chemical or electrocompetent cells. DNA ligases from bacteriophages T3 and T7 are also commercially available, as well as several nonviral DNA ligases, some of which are thermostable. T7 DNA ligase is the enzyme of choice when sticky ends only need to be sealed

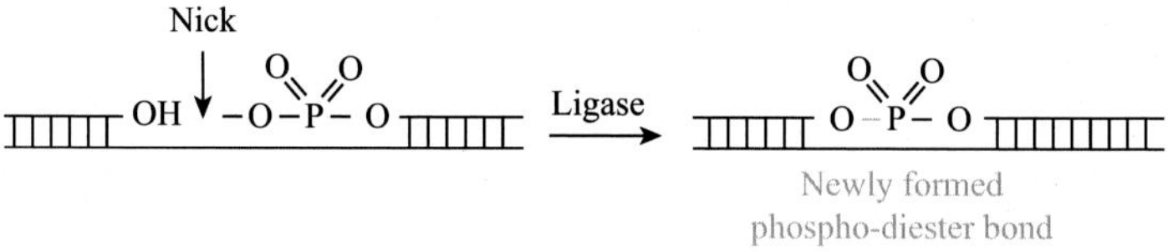

FIGURE 2.14 **DNA ligase overview.** DNA ligases form phosphodiester bond between 3' OH and 5' phosphate of adjacent nucleotides. Most ligases employ ATP as a cofactor and form covalent adenylyl intermediate (not drawn for simplicity), which activates the 5' phosphate of the downstream fragment to be ligated. Nucleophilic attack of the 3' OH at the activated 5'phosphate results in the formation of phosphodiester bond.

since its affinity to sticky ends grossly outweighs the one for blunt ends. T3 ligase is more salt tolerant than T4 and T7 ligases and it is the preferred ligase in applications requiring high ionic strength conditions. The use of DNA ligases in cloning is on decline due to the development of ligation-independent cloning approaches (Fig. 2.9) and recombination-based cloning technologies (Chapter 3), however, their overall value as molecular biology reagents is increasing as more and more techniques employing adapter ligation (for example, SAGE, Fig. 2.7) are being developed. Some of the new-generation sequencing methods also employ T4 DNA ligase to attach adapters to genomic DNA fragments to be sequenced.

Ligase itself drives sequencing by ligation, the newest concept in the fast changing landscape of DNA sequencing. Currently, two major platforms for whole genome sequencing employ the principle: SOLiD (Thermo Fisher, Inc.) and Complete Genomics (Beijing Genomics Institute). The SOLiD technology immobilizes DNA fragments subject to sequencing on beads attached to a slide. Bead/DNA attachment is accomplished via universal adapter. An anchor sequence, complementary to the adapter provides the 5' phosphate group to be ligated to the 3' hydroxyl group of the upcoming probe. Fluorescent sequencing probes consist of two known nucleotides in position 1 and 2 (depicted in color in Fig. 2.15), followed by a stretch of degenerate or universal bases depicted with N and Z. At each cycle only the probe with a perfect match of the two known nucleotides will ligate. The slide is imaged and the unextended molecules are capped by unlabeled probes to maintain cycle synchronization. The fluorophore and the terminal degenerated bases are cleaved off the probe, resulting in net 5 nt extension. The process is repeated 10 times resulting in a string of nucleotides in which two out of every five bases are identified. All ligated probes are then removed, and the entire process of probe binding, ligation, imaging, and cleavage is repeated four times, each with different anchors. At each round the new anchor is offset with one nucleotide allowing the entire sequence to be deciphered. The sequence is assembled computationally based on the imaging snapshots taken after each cycle.

The Complete Genomics technology employs fluorescent probes containing only one known nucleotide and several degenerate nucleotides. The DNA genome of interest is fragmented, cloned with flanking synthetic adapters, and then amplified to form DNA nanoballs. Each DNA nanoball is a long concatemer, product of rolling circle DNA amplification with bacteriophage phi29 DNA pol. Arrays of nanoballs are assembled in a flow cell, which is imaged after every sequencing cycle. Each cycle starts with the hybridization of an anchor sequence to one of the synthetic adapters. Then a pool of fluorescently labeled probes is flown allowing complementary probe hybridization and ligation to take place. The pool of unligated probes is washed away, and the flow cell is imaged. The same steps are performed repeatedly using a set of anchors with offset of one

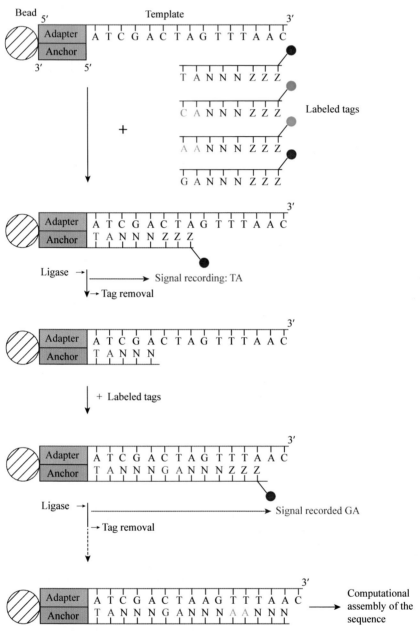

FIGURE 2.15 Sequencing by ligation. Sequence by ligation approach employs T4 DNA ligase to link short fluorescently labeled probes to the 5′ phosphate of an anchor sequence. Ligation takes place only after the known nucleotides in the probe (depicted with color) stably hybridize with the sequenced DNA. Fluorescence is then imaged allowing the outcome to be read. The fluorescent tag is cleaved together with 3 degenerated probe nucleotides (depicted with the letter Z) and the sequencing cycle repeated 10 times, resulting in a DNA read in which every two (in color) out of each 5 nucleotides (2 in color + 3 Ns) are identified. The "N" nucleotides are identified in subsequent rounds with a set of anchors, each with an offset of one nucleotide. The final DNA sequence of the fragment is assembled computationally by processing the image snapshots taken after each sequencing step.

nucleotide, and the DNA sequence is assembled computationally. A separate sequence-specific set of anchors is used for each adapter. Sequences derived from different library clones are assembled into the sequence of the entire genome.

2.3.2.4 Topoisomerases

Topoisomerases are enzymes specializing in removing supercoiling at the replication fork. Depending on their mechanism, they are classified into Class I and Class II based on the number of cuts being introduced in their substrate. Class I enzymes introduce one cut, assist in strand rotation to remove supercoiling and then reseal the cut (Fig. 2.16). Class II enzymes accomplish the same outcome by introducing cuts in both DNA strands. Vaccinia virus topoisomerase I is used in the TOPO cloning technology, which allows fast and efficient cloning in a single reaction performed at room temperature. TOPO cloning kits for TA cloning, blunt, end cloning, and directional cloning are available. Vectors permissive to TOPO cloning contain Vaccinia topoisomerase recognition sequence (5′C/TCCTT3′). The enzyme cleaves the sugar-phosphate backbone and forms a covalent intermediate involving tyrosine residue from its active site (depicted with

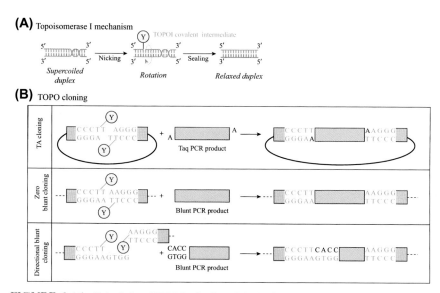

FIGURE 2.16 **Principle of TOPO cloning.** TOPO cloning technology is based on the biochemical activity of vaccinia virus topoisomerase I. The enzyme recognizes the 5′-(C/T) CCTT-3′ target sequence, introduces a single cut allowing DNA to unwind, and then reseals the sugar-phosphate backbone effectively removing DNA supercoiling. The employed mechanism involves a covalent enzyme–substrate intermediate between the cleaved DNA strand and a tyrosine residue depicted as a circle marked with Y (A). TOPO vectors allowing TA-, blunt-, or directional blunt cloning have been developed (B).

a blue circle labeled with Y). Once a DNA fragment with a cognate end is provided in the reaction, the enzyme joins the ends of the vector and the fragment generating covalently closed dsDNA molecules which are transformed in competent cells and subjected to further analysis. A hallmark of the TOPO cloning technology is the very high efficiency (>95%). TOPO cloning vectors for different purposes (cloning for sequencing, in vitro transcription, long fragment cloning, fragment subcloning) are commercially available.

2.3.2.5 Single–Stranded DNA Proteins

Single-stranded DNA proteins (SSBs) bind ssDNA nonspecifically and are involved in all cellular pathways involving DNA manipulations. It is thought that their primary role is to protect ssDNA from nucleases as it arises during DNA replication, recombination, and repair. Surprisingly, many viruses, both prokaryotic and eukaryotic, code for their own SSBs (usually essential gene for replication) despite the fact that the host cell has abundance of SSB itself. SSBs interact with other players at the replication fork and were shown to coordinate the leading and lagging strand of DNA synthesis most likely by "organizing" the replisome via multiple protein–protein interactions. Bacteriophage T4 SSB, gp32, is commercially available and commonly used as an agent for removal of DNA secondary structure when DNA templates prone to secondary structure formation are amplified by PCR, as well as in biochemical assay. In electron microscopy, T4 SSB is utilized as a tool to "mark" ssDNA regions.

2.3.3 DNA-Modifying Enzymes

2.3.3.1 Polynucleotide Kinase

Polynucleotide kinase (PNK) is an enzyme that catalyzes the reversible phosphorylation of nucleoside monophosphates, ss and ds nucleic acids. PNK from bacteriophage T4 is widely used in molecular biology for 5′ phosphorylation of DNA and RNA for the purpose of labeling and ligation, as well as the removal of 3′phosphoryl groups from various nucleotide and nucleic acid substrates for labeling purposes. T4 PNK works by transferring the y phosphate of ATP to the 5′ end of nucleic acids. Genetically, engineered version lacking phosphatase activity is also available.

2.3.3.2 Methyltransferases

Methyltransferases are enzymes that transfer methyl groups to their substrates resulting in substrate methylation. Virus-related nucleic acid methyltransferases (MTases) are described as MTase components of the restriction modification systems in bacteria or are individually coded by viral genomes. Methyltransferase applications are relevant to epigenetic studies and analysis of the nucleic acid modifications. MTase components of restriction modification systems work in a manner similar to the REs; however, they methylate their recognition sequences or sequences in close vicinities instead of cleaving the sugar-phosphate backbone. The applications of these MTases are limited to targeted modification of their recognition sequences. The GpC methyltransferase M.CviPI, coded by the M.CviPI gene of *Paramecium bursaria Chlorella* virus NYs1, methylates cytosine in the context of 5'GpC3' sequence using S-adenosyl methionine (SAM) as a methyl donor. The enzyme is used as a tool to interfere with REs cleavage, to uniformly label DNA with tritium (^{3}H), as well as to alter the physical properties of DNA. Methylated cytosines have been shown to lower the free energy of transition to Z-DNA and to increase the helical pitch of the double helix, thus allowing controlled manipulation of the biophysical characteristics of DNA. Bacteriophage T4 beta-glycosyltransferase is another viral enzyme related to methylation studies. The enzyme transfers glucose from uridine diphosphoglucose (UDP-glucose) to methylated hydroxymethylcytosines found in dsDNA molecules. It is thought that hydroxymethylcytosine is a product of methylcytosine oxidation. Its role is not well understood and different scenarios have been proposed for different systems. In bacteriophages, it is thought that the presence of hydroxymethylcytosine is an evolutionary adaptation to evade the restriction modification systems of bacteria. In humans it has been found in stem cells and brain, where it supposedly influences gene regulation directly or indirectly by promoting demethylation. The T4 beta-glucosyltransferase is used for detection of hydroxymethylcytosine and radioactively labeling hydroxymethylcytosine moieties of DNA. It is also utilized in the context of the HELP assay (described in Section 2.3.1.2) as means to quantify hydroxymethyl and methylcytosine content of DNA. The MspI RE used in that assay is not sensitive to methylation and will cut both methylated and unmethylated DNA; however, if DNA is treated with beta-glycosyltransferase the hydroxymethylcytosines will be glycosylated preventing MspI cleavage. The parallel analysis of DNA with HpaII (sensitive to methylation), MspI (insensitive to methylation) and combination of beta-glycosyltransferase treatment and MspI cleavage (sensitive to glycosylated hydroxymethylcytosine) allows for quantification of methylation levels and differentiation between methyl and hydroxymethyl modifications.

2.4 VIRAL TOOLS FOR MANIPULATING RNA IN VITRO

2.4.1 RNA Polymerases

By their nature RNAPs are DNA-dependent RNA polymerases (RNAPs) that utilize a DNA template to synthesize RNA molecules. Bacteriophage

RNAPs are an essential tool in molecular biology when it comes to RNA biology and protein expression. Advances in RNA biology in the fields of general RNA structure and structure–function analyses, splicing, translation, ribozymes, RNAi, RNA–protein interactions, RNA viral infectivity, and virtually any other area of modern biology involving RNA molecules would not be possible without in vitro RNA synthesis driven by bacteriophage polymerases. While methods for chemical synthesis of RNA have been developed, their utility is very limited due to high cost and inability to sustain synthesis of long molecules. In addition to RNA studies, RNA is synthesized in vitro for the purpose of in vitro translation, as a labeled probe and expression control via antisense RNA. Virtually any protein that needs to be purified is attempted to be expressed in *E. coli* first, and the production of relevant mRNA is accomplished by cloning the gene of interest under the control of bacteriophage promoter and introducing it in *E. coli* for in vivo protein expression. Viral tools for protein expression are discussed in detail in Chapter 4.

Bacteriophage RNAPs are preferred enzymes for in vitro transcription since they are comprised of single polypeptide that executes all aspects of transcription as opposed to the highly complex bacterial and eukaryotic RNAPs, which employ multiple subunits. Historically, three different bacteriophage RNAPs have been utilized in molecular biology originating from bacteriophage T3, bacteriophage T7, and bacteriophage SP6, among which the T7 system is preferentially used. T7RNAP recognizes its cognate promoters with very high specificity initiating efficient transcription that outpaces the cellular one by half of an order of magnitude. Since the sequence of the T7 promoter differs significantly from the one for *E. coli* RNAP, the T7 promoter/RNAP/terminator system is very efficient and can transcribe essentially any sequence placed in the right context between the two regulatory signals. In fact, T7RNAP promoters are grouped into two classes that slightly differ from each other. Class II promoters are found in genes relevant to DNA synthesis, expressed 6–15 min postinfection, whereas Class III promoters are found in front of genes relevant to packaging and capsid formation, expressed 7 min postinfection until lysis. The standard T7 promoter consensus used in molecular biology is 5′ TAATACGACTCACTATAG 3′ corresponding to Class III promoters. Unlike the *E. coli* RNAP, T7RNAP is not sensitive to rifampicin, thus allowing the genes cloned under the control of the T7 promoter to be exclusively transcribed in the presence of the antibiotic. T7, T3, and SP6 RNAPs are structurally similar and execute their polymerase function by a DNApol I-like polymerase domain, comprised of the typical palm, thumb, and fingers' structural elements; however, unlike DNA pol I they can initiate de novo synthesis without a primer and do not have a proofreading activity. Mitochondrial RNAPs belong to the same group; however, they require sigma-like subunit for promoter recognition. In addition to the polymerase domain, T7RNAP has several accessory domains that contribute to promoter recognition (recognition loop); promoter opening and enzyme processivity (N-terminal domain); and regulation by lysozyme (C-terminal domain). The specificity of T7RNAP can be changed if the domain responsible for promoter recognition is swapped with the one of another RNAP from the same group. Specificity change was also achieved in a proof of principle experiments utilizing phage-assisted continuous evolution, a technique based on phage display, which is described in detail in Chapter 5. T7RNAP activity is regulated by lysozyme, the T7 enzyme lysing the cell wall on the release of the phage progeny. Lysozyme interacts with the C-terminal domain of the RNAP inflicting a conformational change, which lowers the affinity of the molecule for NTPs favoring preferential expression from the Class III promoter, which can be initiated at lower NTP concentration. Transcription initiation is a stepwise process including (1) promoter binding, (2) promoter melting, (3) promoter release (preceded by 9 nt RNA synthesis), and (4) RNA extension of initial transcripts (up to 14 nts) to allow the formation of stable elongation complex. The transition between initiation and elongation state of the RNAP takes place when the transcript length is in the range of 9–14 nts, after which synthesis proceeds with high speed until terminator sequences are encountered. Two variants of T7RNAP terminators have been described, designated as Class I, similar to rho-dependent terminators in *E. coli*, which employ hairpin as a structural element and Class II terminators that do not form any secondary structure and are somewhat weaker. Molecular cloning and in vitro transcription protocols utilize Class I terminators. Shorter versions of Class II terminator function as pausing site and are found in several locations throughout the T7 genome including near the end of the genome, effectively causing the RNAP to pause near the concatemeric junction in the long genomic concatemers produced in phage replication. It is thought that the position of the pausing sequence near the concatemeric junction has been selected in evolution to prevent interference between transcription and genome packaging. Similarly, promoter sequences positioned near the origin of replication have been considered an evolutionary adaptation positively influencing replication initiation. The high promoter specificity and fast rates of synthesis make T7RNAP an ideal choice for in vitro transcription applications. Mutant version of the enzyme has been engineered to incorporate modified nucleotides and mix of ribo- and deoxynucleotides, enabling synthesis of modified RNA molecules for various structural studies and synthesis of RNAse-resistant RNA molecules for the purpose of generating RNA therapeutics with increased half-life. T7RNAP has been and continues to be the enzyme of choice for understanding the mechanisms of RNA transcription as well as the system for proof of principle experiments when new technologies are being developed.

Most in vitro transcription vectors include bacteriophage promoters on each site of the cloned gene, allowing for production of sense and antisense RNA. Generally, T7 and T3 RNAPs yield similar amounts of RNA, whereas SP6 yields are slightly lower. RNAP from the marine phage Syn5 is actively researched as a potential new addition to the group of bacteriophage polymerases used for in vitro transcription. Commercial kits for in vitro transcription on different scales are available, as well as kits for in vitro transcription/translation utilizing lysates from different types of organisms (E. coli, wheat germ extract, rabbit reticulocytes). Bacteriophage RNAPs work efficiently in eukaryotic systems in vivo and in vitro if the cognate promoter/terminator sequences are present. Various DNA templates (linear DNA fragments, circular plasmids, PCR products, cDNA, in vitro annealed oligonucleotides can be used as templates for in vitro transcription). To be translated eukaryotic mRNAs need to have their ends properly modified. The 5′ end of most eukaryotic mRNA is capped, whereas the 3′ end is polyadenylated. Both modifications increase stability of mRNA and ensure compatibility with the eukaryotic translation machinery. Polyadenylated transcripts have been described in bacteria, chloroplasts, and mictochondria; however, in these systems the addition of a polyA tail destabilizes transcripts. The existence of E. coli polyA polymerase has been known to science for a long time; in fact the enzyme was initially isolated before the significance of polyadenylation in eukaryotes was recognized. E. coli polyA polymerase I is a single polypeptide that can add ribonucleotides to preexisting 3′OH end of RNA molecule in a template-independent manner, although its affinity for different ribonucleotides varies. The purified enzyme readily adds polyA tails to purified transcripts produced by in vitro transcription. RNA capping has proved to be a more challenging process. Initially, synthetic CAP structures (a guanine nucleotide methylated at position 7 and connected to mRNA via an unconventional 5′–5′ triphosphate linkage) were synthesized and used in huge excess during initiation of in vitro transcription. The process yielded mixes of capped and uncapped mRNAs, and only a fraction of the capped mRNAs had the methylguanine inserted in the correct orientation. Later modification of the synthetic CAP was developed that supported only the chemistry of the correct CAP orientation. Synthetic CAP is essentially a G dinucleotide in which both monomers are connected with 5′–5′ triphosphate bond, and the second one is methylated in position 7. Adenine CAPs (A CAPs), in which the unmodified G is replaced with A, have been synthesized and can be added to the 5′ end of RNA molecules to increase their stability for transfection and microinjection experiments where translation is not needed since the A CAP renders the RNA molecule incompatible with the ribosome. The introduction of vaccinia virus capping system allowed for 100% efficient capping in in vitro transcription reactions.

2.4.2 Vaccinia Virus RNA Capping System

Fig. 2.17 depicts the process of in vitro mRNA capping utilizing the vaccinia virus system. The system employs two enzymes: one introducing the initial CAP structure, known as CAP 0, and a second enzyme, mRNA CAP2'-O-Methyltransferase converting CAP 0 to CAP 1 structure. The capping enzyme is a two subunit protein exhibiting three enzyme activities: RNA triphosphatase, guanylyltransferase (both housed on the D1 subunit), and guanine methyltransferase (housed on the D2 subunit). The coordinated action of the all three activities results in the synthesis of CAP 0. mRNA CAP2'-O-Methyltransferase transfers a methyl group from SAM to the 2' position of the first nucleotide connected to the CAP, creating CAP 1 and completing the capping process. In addition to modifying mRNAs for the purpose of translation the vaccinia capping enzyme can be used for 5' labeling of mRNA.

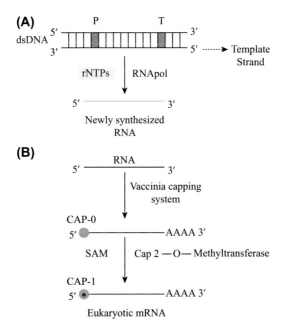

FIGURE 2.17 **In vitro transcription and eukaryotic mRNA capping.** (A) In vitro transcription. In vitro transcription is driven by robust bacteriophage RNAP, which recognizes its cognate promoter (P) and initiates RNA synthesis. Transcription proceeds until a terminator (T) signal is reached, resulting in the release of the newly RNA transcript (please note that RNA and the gene coding it are not drawn to scale). (B) In vitro capping. To be translated eukaryotic mRNAs need to acquire a 3' polyA tail and 5'CAP. The polyA tail is introduced by polyA polymerase (not drawn for simplicity). The CAP is synthesized by vaccinia virus capping system in a two-step process beginning with the addition of CAP 0, depicted with *orange circle*, which is subsequently methylated (depicted with *asterisks*). PolyA and the CAP increase the stability of mRNAs and ensure compatibility with the protein synthesis machinery.

2.4.3 RNA Ligases

So far biology has described limited number of RNA ligases and not surprisingly some of them are coded by viruses. Bacteriophage T4 codes for RNA ligase 1 and RNA ligase 2 ligating ssRNA and dsRNA molecules, respectively. In vivo RNA ligase 1 repairs a break in the anticodon loop of the host tRNA^{Lys} introduced by an *E. coli* encoded nuclease, whose expression is induced by the phage infection. The in vivo significance of RNA ligase 2 for the T4 biology is not well understood. The enzyme belongs to a family of dsRNA ligases, which includes ones involved in RNA editing in parasites such as *Trypanosoma* and *Leishmania*. RNA ligase 1–like proteins have been identified in fungi, baculoviruses, archea, and archeal viruses (thermostable homologues), whereas RNA ligase 2–like enzymes have been identified in *Vibrio* phage KVP40, baculoviruses and entomopoxviruses, some parasites, and archaea species.

T4 RNA Ligase 1 ligates a 5′-phosphoryl nucleic acid donor to a 3′-hydroxyl nucleic acid acceptor by connecting them with a phosphodiester bond. It utilizes ssRNA, ssDNA, and dinucleoside pyrophosphates as substrates and is dependent on ATP hydrolysis for its function. T4 RNA ligase 1 is used for ligation of ssRNA and DNA, RNA ligase–mediated rapid amplification of cDNA ends (RLM-RACE), ligation of oligonucleotide adapters to cDNA or single-stranded primer extension products for PCR, oligonucleotide synthesis, and various 5′ nucleotide modifications of nucleic acids. T4 RNA Ligase 2 ligates adjacent 5′ phosphate at the end of RNA and DNA strands to 3′ OH of RNA strands in the context of nicked dsRNA or RNA/DNA hybrids. The enzyme exhibits significant homology to DNA ligases and mRNA capping enzymes. T4 RNA ligase is used to seal nicks in dsRNA and dsRNA/DNA hybrids. Genetically engineered version harboring K277Q mutation ligates with minimum amount of undesired products such as concatemers. A C-terminal truncated version of T4 RNA ligase 2 lacks the ability to hydrolyze ATP and adenylate 5′phosphates; however, it is able to catalyze ligation with preadenylated products. The truncated version is used for ligating adenylated adapters to mRNAs, thus facilitating their recovery, cloning and characterization, as well as ligation of adenylated primers to RNA for the purpose of creating cDNA libraries of small RNAs and strand specific cDNA libraries.

Viral enzymes and nonenzyme components were among the first molecular biology tools for genome manipulations in vitro, and their number and utility grew over the years. As molecular biology continues to advance marine bacteriophages and viruses of organisms living at extreme conditions are viewed as the next big source of new virus-based tools. Enzymes already in use are being genetically modified or subjected to molecular evolution to select for specific properties or variations of properties that fit perfectly the needs of the technique and/or associated instrumentation. Viral enzymes are also being applied to new fields such as development of rapid sequence-specific DNA amplification for pathogen diagnostics (DNA pols), nanotechnology (phi29 polymerase), vaccines (RNA pols), among others.

Further Reading

1. Chan SH, Stoddard BL, Xu SY. Natural and engineered nicking endonucleases–from cleavage mechanism to engineering of strand-specificity. *Nucleic Acids Res* 2011;**39**(1):1–18.
2. Chen CY. DNA polymerases drive DNA sequencing-by-synthesis technologies: both past and present. *Front Microbiol* 2014;**5**:305.
3. Goodwin S, McPherson JD, McCombie WR. Coming of age: ten years of next-generation sequencing technologies. *Nat Rev Genet* 2016;**17**(6):333–51.
4. Nichols NM, Tabor S, McReynolds LA. RNA ligases. *Curr Protoc Mol Biol* 2008. Chapter 3:Unit3. 15.
5. Patron NJ. DNA assembly for plant biology: techniques and tools. *Curr Opin Plant Biol* 2014;**19**:14–9.
6. Pingoud A, Wilson GG, Wende W. Type II restriction endonucleases–a historical perspective and more. *Nucleic Acids Res* 2014;**42**(12):7489–527.
7. Tabor S, Richardson CC. A single residue in DNA polymerases of the *Escherichia coli* DNA polymerase I family is critical for distinguishing between deoxy- and dideoxyribonucleotides. *Proc Natl Acad Sci USA* 1995;**92**(14):6339–43.
8. Vasu K, Nagaraja V. Diverse functions of restriction-modification systems in addition to cellular defense. *Microbiol Mol Biol Rev* 2013;**77**(1):53–72.
9. Zhu B. Bacteriophage T7 DNA polymerase – sequenase. *Front Microbiol* 2014;**5**:181.

Viral Tools for Genome Manipulations In Vivo

The molecular biology revolution advanced dramatically our understanding of the molecular basis of living matter and the technical capabilities of science to manipulate nucleic acids in vitro. It also created unprecedented ways to manipulate the genomes of bacteria, animals, plants, and viruses in vivo thus paving the path science discoveries to be translated into practical solutions of current challenges in medicine, agriculture, and biotechnology. Ever since the first discoveries of natural mechanisms of gene transfer (transformation, conjugation, and transduction) at the dawn of molecular biology, the deliberate manipulations of genomes have been used as a tool to understand heredity, identify genes and their functions, as well as to decipher the mechanisms of the flow of genetic information. Essentially, genome manipulations are continuing to do exactly that today, however, on much larger scale and with more precise tools delivering results much faster with less specialized effort. Not surprisingly viruses, viral components, and components of the cellular antiviral defenses in prokaryotes and eukaryotes have been routinely used as tools in genome manipulations. These will be discussed in the context of conventional approaches to genome manipulations, recombineering, and gene editing.

3.1 CONVENTIONAL APPROACHES TO GENOME MANIPULATIONS

Until modern tools of molecular biology were developed, science had very limited capabilities to alter genomes, and most genome manipulations outside classical crosses took place in bacteria utilizing the natural mechanisms of gene transfer: conjugation, transformation, and transduction. From those only transduction is directly mediated by viruses, although phages are instrumental in releasing bacterial genomic DNA and plasmids in the environment that may transform bacterial cells surrounding them (Fig. 3.1). Transduction is the process of transferring DNA between two

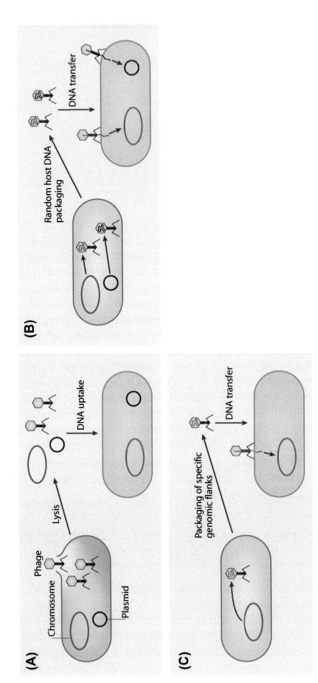

FIGURE 3.1 **Natural processes of genetic material exchange mediated by viruses.** (A) Transformation. Viruses contribute indirectly to transformation by releasing bacterial chromosomal and plasmid DNA to the environment when lysing their host cells. (B) Generalized transduction. Lytic and lysogenic viruses can package random pieces of host chromosomal genetic material in their capsids and transfer it to the next host. (C) Specialized transduction. When excising themselves from host genomes, lysogenic viruses occasionally package genes flanking their insertion site. *Reprinted from Salmond GP, Fineran PC. A century of the phage: past, present and future. Nat Rev Microbiol 2015;13(12):777–86, with permission from Nature publishing group.*

cells, donor and recipient, utilizing a virus as a "transport carrier." The natural frequency of transduction is very low and virus specific. Transduction takes place by two distinct mechanisms: generalized and specialized. Generalized transduction is described in bacteriophages that package their genome by head-full mechanism such as bacteriophage P1 infecting *Escherichia coli* or bacteriophage P22 infecting *Salmonella*. Both P1 and P22 are lysogenic phages; however, P1 establishes lysogeny by persisting the cytoplasm as an episome, whereas P22 integrates in the host genome. Pieces of host DNA are occasionally packaged by error and transferred to the next cell during subsequent infection. Once the donor bacterial DNA is delivered in the cytoplasm of the recipient cell, it has a chance to be incorporated in its genome by homologous recombination (HR). If a recombination takes place successfully, the donor genetic material is permanently inserted into the bacterial chromosome of the recipient. If recombination does not take place, the donor DNA is degraded. If appropriate markers are available, transduced cells can be identified and analyzed by plating on various selective media. Conclusions regarding the distance between genes can be used to generate low-resolution genetic maps based on the frequency of transduction of pairs/groups of genes/selectable markers. Frequent transduction signifies close distance, whereas infrequent transduction signifies larger distance between genes. In the age of sequencing, genetic engineering, and bioinformatics the approach appears largely cumbersome and inefficient, however, in the past it has contributed greatly to understanding the genome organization of key prokaryotic model systems.

Specialized transduction refers to a process in which only specific genes can be moved from cell to cell. Bacteriophage *lambda*, for example, can transduce only the DNA sequences flanking its integration site (genes *bio* and *gal*), essentially as a result of erroneous excision process. Lambda transduction is used as a tool to transfer genetic markers between *E. coli* strains. For example, if a *gal⁻ lambda* nonlysogenic strain is mixed with a wild-type *lambda* lysate and then plated on selective media lacking *gal*, only cells that have acquired *gal* by transduction will grow. The newly generated strain with a *gal* marker could be subsequently used in genetic experiments. In certain experimental conditions, phages can transduce other phages. For example, bacteriophage *T1* grown in *lamda* lysogenic strains can transduce lambda genome to new host cells. The frequency of transduction is greater for multilysogenic *E. coli* strains and can be enhanced by overexpression of *Lambda Red* recombination genes or the *E. coli recE* genes.

Today transduction of bacterial cells is mainly used to deliver DNA of interest to bacterial hosts. Experimental systems for transduction of genes of interest in animal and plant cells were developed much later and currently present powerful tolls for gene delivery in research, protein expression, gene therapy, etc.

In laboratory environment, the genomes of mammalian cells are modified by transfection. The original term, a hybrid of transformation and infection, was introduced to describe the transformation of infectious viral DNA into bacteria resulting into infectious viral particles. Since the term transformation in animal cell was already in use referring to malignant transformation in cancer, the term transfection was adopted to describe the process of purposeful introduction of naked/purified genetic material into animal cells. Conceptually, on the mechanistic level transformation in bacterial cells and transfection of animal cells describe the same process. As the methods for transfection evolved over time and lentivirus-based transduction system was developed as method to deliver nucleic acids in mammalian cells, the meaning of the term evolved from mechanistic to cellular level, now referring to modification of cellular properties as a result of introduction of nucleic acids (Fig. 3.2). Both DNA and RNA can be introduced in animal cells by transfection. DNA is usually introduced as a plasmid and can persist in the cell transiently or to physically integrate into the cellular genome resulting into stable transfection. RNA persists in the cell only transiently, and it is usually transfected in nondividing cells for the purpose of gene expression and RNA decay studies, as well as a part of RNA-silencing protocols modulating gene expression. Transfection of RNA instead of DNA is often used as an approach to limit the persistence of a protein in the cell or as a safeguard against accidental integration of transfected DNA. Both phenomena are of key importance in recombineering and gene editing technologies as they are applied to agriculture and medicine. Introduction of nucleic acids into mammalian cells can be achieved by multiple approaches such as coating nucleic acids

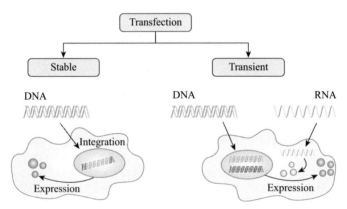

FIGURE 3.2 **Transfection of mammalian cells.** Transfection is the process of deliberate introduction of purified nucleic acids into mammalian cells growing in tissue culture. Stable transfection is associated with physical integration of the transfected nucleic acid in the cellular genome. Stably transfected cell lines are continuously maintained/stored in the lab and repeatedly used in experiments pertinent to the inserted genetic material. Transient transfection is associated with temporary persistence of the introduced genetic material and is performed de novo for each subsequent experiment.

with calcium phosphate, embedding nucleic acids in liposomes, viruses and virus-like particles, electroporation, microinjection, among others. Transduction-based gene delivery takes advantage of the natural propensity of a virus to deliver capsid-packaged nucleic acid inside a host cell. Usually, the gene(s) of interest is/are packaged into replication-deficient viral particles. Retroviruses (including lentiviruses), adenoviruses, adeno-associated viruses, and herpesviruses are most frequently utilized. Although powerful, transduction is not always preferred method for introduction of nucleic acids in mammalian cells due to cost, small size limits and risks of cytopathic effects, random integration, and gene disruption. Stably transfected cell lines are used as packaging factories to produce large quantities of transducing viruses carrying nucleic acids of interest.

The genomes of the plant cells are generally manipulated by transformation utilizing the properties of the *Ti* plasmid of *Agrobacterium tumefaciens*. Recently, the laboratory of George Lomonossoff (John Innes Centre for Plant Biotechnology, UK) developed a cowpea mosaic virus–based system for transient expression of heterologous proteins in plants. The system combines Ti transformation and deconstructed viral vectors allowing for fast and efficient delivery of genes of interest to the entire plant and high yields of protein production. The new approach is the basis of the rapidly advancing field of molecular farming, which holds great promise for production of vaccines and protein-based drugs, as well as industrially significant metabolites and enzymes.

Baculoviruses are widely used to deliver genes of interest in insect cells for the purpose protein expression and purification and are currently emerging as a promising delivery tool for mammalian cells.

3.2 VIRUS-BASED RECOMBINATION SYSTEMS AND THEIR APPLICATIONS

3.2.1 Virus-Driven Recombination Overview

Genetic recombination is a driving force of diversity in nature, both on the organismal and cellular level. Although not exactly living entities, viruses are no strangers to recombination processes and benefit from their outcomes in the constant tug-of-war with their host. For some viruses, recombination presents a key component of their genome replication strategies (recombination-mediated replication in bacteriophage T4 and herpesviruses, for example), for others it is an essential mechanism to insert themselves in the host genome (for example, the bacteriophage lambda integration/excision machinery) or to separate the replication products of the lysogenic episome (for example, Cre recombinase of bacteriophage P1). In addition, recombination events in the context of natural selection result into acquisition of new mutations and characteristics giving viruses advantage in evolution (for example, the influenza virus reassortment) or in reconstituting active viruses from complementing viral genomes harboring detrimental mutations. In the world of molecular biology and genetic engineering, viral recombination played a critical role in establishing approaches to reconstitute viruses in laboratory settings by transfection of infectious DNA and to push the size limits of DNA fragments that can be manipulated.

Recombineering is a relatively young field, defined as recombination-mediated genetic engineering, which allowed for precise manipulations of large DNA fragments in vivo beyond the capacity of in vitro genetic engineering techniques. Recombineering achieves essentially the same goals as restriction enzyme–based in vitro cloning techniques: (1) it can replace genetic fragments or can be used to generate insertions and deletions; (2) it can be used for subcloning; or (3) it can be used to introduce of point mutations (Fig. 3.3). The approach is largely based on the properties of the *Lambda Red* and *Lambda Red*–similar recombination systems

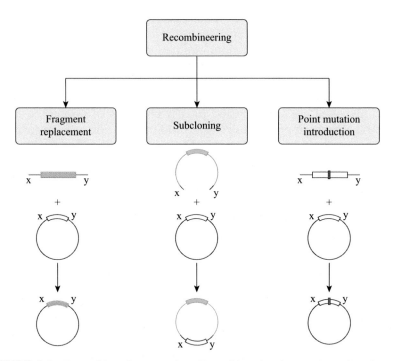

FIGURE 3.3 **Recombineering overview.** Recombineering is genetic engineering performed with recombination-based tools. Conceptually, recombineering can be used to replace fragments, thus completely changing the sequence of the genetic material between the regions of homology (designated with x and y); for subcloning and introduction of point mutations. Deletions and insertions are generated when the replacement fragment is shorter or longer than the original one. Recombineering approaches based on both homologous and site-specific recombination have been developed.

employing a pair of $5' \rightarrow 3'$ exonuclease and ssDNA-binding protein with strong annealing activity to mediate recombination of sequences sharing limited homology. Site-specific viral recombinases such as the Cre/LoxP system of bacteriophage P1 have been used for target specific genome manipulations, whereas the lambda excision/insertion machinery has been utilized in the recombination-based Gateway cloning system.

3.2.2 *Lambda Red* Recombineering

Lambda Red recombination system consists of three protein components: Red alpha, a $5' \rightarrow 3'$ nuclease coded by the *exo* gene; Red beta, an ssDNA-binding protein with annealing activity coded by the *bet* gene; and Red gamma, an inhibitor of the host RecBCD nuclease coded by the *gam* gene. The genes for the Red system are coded by the P_L operon, which is expressed early in the phage life cycle. Recombination events driven by the Red system utilize double-stranded breaks in DNA and proceed via two principle mechanisms: annealing and strand invasion (Fig. 3.4). The annealing mechanism takes place when both homologous partners

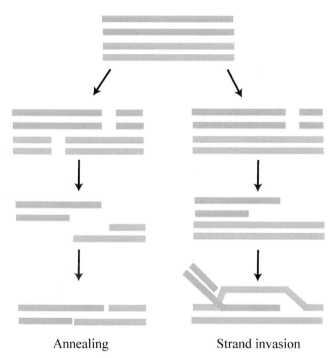

Annealing   Strand invasion

FIGURE 3.4   **Conceptual mechanism of *Lambda Red* recombination system.** The *Lambda Red* system for homologous recombination utilizes dsDNA breaks and repairs them via an annealing mechanism when both substrate molecules are broken (**left**) or via strand invasion mechanism when only one of the substrate molecules is broken (**right**). The strand invasion pathway is RecA dependent.

contain double-stranded DNA breaks. The Red exonuclease creates 3′ overhangs by selectively digesting one of the DNA strands, homologous regions anneal with the assistance of the Red beta protein, and the resulting DNA gaps are repaired. Strand invasion mechanism takes place when only one of the homologous partners is broken. In this case the Red alpha generated 3′ overhang invades the homologous dsDNA with the help of Red beta and RecA. In both mechanisms, the Red gamma protein assists the process by preventing the action of the RecBCD helicase/nuclease complex, which is known to load on dsDNA breaks and digest both strands until it reaches a Chi (crossover hotspot instigator) site initiating host recombination-based repair. Red gamma effectively shuts down the host recombination and ensures that the Red system is working preferentially. The significance of the Red system for the biology of the *lambda* bacteriophage is not completely understood. Deletion of the *exo* and *bet* genes interferes with the phage growth in laboratory conditions rather mildly. It is thought that the Red system may stimulate phage replication by circulating concatemer genomes, which present better replication templates and/or function to quickly repair dsDNA breaks inflicted by the host restriction modification systems when the phage is infecting new cells. It is possible that in nature lambda replicates in conditions triggering high rates of DNA damage, which makes a recombination system a necessity. In the laboratory, the Red system (expressed from the bacterial chromosome or a plasmid) promotes HR between the host chromosome and linear dsDNA fragments introduced by electroporation resulting in gene replacement. It also supports recombination of short ssDNA oligonucleotides introduced in the cell with the bacterial chromosome resulting in gene conversion, i.e., unidirectional transfer of genetic material from the oligonucleotide donor to the chromosome acceptor. The latter function requires only the Red beta protein. The *lambda* proteins *Rap* and *Orf* have been associated with the Red system and thought to function as nonessential accessory factors helping to "focus" the recombination activity at the cos sequences at the ends of the linear phage genome and promote its circularization. The Red system is not restricted to *E. coli* and has been shown to work efficiently in other bacterial species such as *Salmonella*, *Klebsiella*, *Mycobacterium*; however, its efficiency decreases in correlation with the evolutionary distance of the bacteria to its natural host. Furthermore, the system efficiency can be improved if larger homologous regions are employed. *Lambda Red*–like recombination systems have been described in the *Rac* prophage of *E. coli* and the *P22* phage of *Salmonella*. The *Rac* phage codes for RecE and RecT recombination proteins, corresponding to the Red exonuclease and Red ssDNA–annealing proteins, respectively and thought to support recombination essentially by the same mechanisms. The RecE/RecT system supports efficient recombination of linear dsDNA fragments with limited homology (only 42 nt), which allows homology sequences to be easily included in PCR primers and recombination to be carried out with electroporated PCR fragments. No accessory proteins have been associated with the RecE/*RecT* system.

The *P22/Salmonella* system employs four proteins (ssDNA-annealing protein, Erf; an anti-RecBCD protein, Abc2, and two accessory factors) working together to circularize the repeats at the end of the linear dsDNA genome before replication can take place. Interestingly, the Abc2 protein interacts with RecBCD and modifies its nuclease activity forcing it to work together with the *P22* ssDNA-annealing protein. Unlike *Lambda Red*, the *P22* recombination system is essential for phage replication.

In vivo recombineering is most efficient when a brief high-concentration "pulse" of Red proteins is "supplied" into the cell environment. Usually, this is accomplished by their expression from plasmids under the control of arabinose-inducible promoter or from a *Lambda* prophage under the control of the natural transcription regulatory system of the bacteriophage modified to include temperature-sensitive lambda repressor. Short temperature shifts allow repression to be removed for a controllable period of time resulting in very limited expression of Red proteins, thus promoting efficient recombineering without potential toxic effects of Gam overexpression or background recombination between genome areas harboring repeats.

Recombineering is used to manipulate plasmids, bacterial chromosomes and bacterial artificial chromosome (BAC) vectors with the goal to engineer insertion, deletion, or point mutations most frequently applied to addition/removal of selectable or nonselectable markers, tags, introduction of site-specific recombinases and/or their target sites, tissue-specific promoters, among others. Molecules manipulated by recombineering are most frequently utilized to generate transgenic plants and animals, manipulate embryonic stem cells or for the purpose of gene therapy. Major advantages of the approach are as follows: (1) fast and user-friendly procedure effectively replacing long and cumbersome protocols for restriction enzyme-based manipulations in vitro (Fig. 3.5) and (2) the large size limits of the manipulated DNA molecules, which makes the technology valuable for engineering of eukaryotic genome fragments cloned on BACs. The short length of homologous regions required recombination to take place can be perceived as both advantage and disadvantage, allowing easy design of substrates for recombineering and requiring strict expression of Red enzymes to limit undesired recombination between regions with repeats, respectively. Unfortunately, Red-like recombination systems of bacteriophage origin are not active in eukaryotic cells; however, Red-like recombination systems have been reported in *Herpes simplex virus type 1* (UL12, alkaline nuclease with $5' \rightarrow 3'$ exonuclease activity, and ICP8, a ssDNA-binding protein with annealing activity) and in *Autographa californica multinucleocapsid*

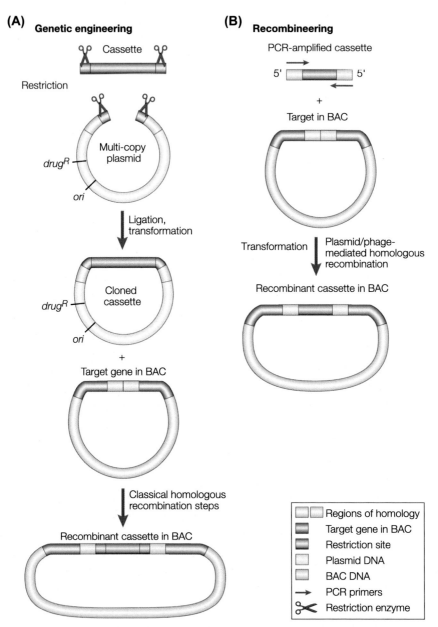

FIGURE 3.5 **Classical genetic engineering approaches versus recombineering.** Genetic engineering by recombination, recombineering, can accomplish essentially the same experimental goals, however, much faster and in smaller number of steps in comparison with classical genetic engineering approaches employing restriction enzyme–based DNA manipulations in vitro. *Reprinted from Copeland NG, Jenkins NA, Court DL. Recombineering: a powerful new tool for mouse functional genomics. Nat Rev Genet 2001;2(10):769–79, with permission from Nature publishing group.*

nuclear polyhedrosis virus (AN/Lef-3 alkaline nuclease/ssDNA-binding protein pair). The herpes system is relatively well characterized and could provide a valuable starting point for development of mammalian recombineering approach. The utility of the *Lambda Red* and *Rac RecE/ RecT*-based recombineering methodologies triggered database analyses aiming to identify similar systems in more phages. The efforts resulted in the discovery of similar recombination systems in bacteriophages and prophages of *Mycobacterium, Listeria, Legionella*, among others which are hoped to contribute to genome manipulations of these pathogens for the purpose of better understanding their biology and future biomedical applications. In addition, ssDNA oligo-based recombineering approaches employing the concept of the *Lambda Red* system have been developed in yeast (YOGE—yeast oligo-mediated genome engineering) and lactic acid bacteria utilizing overexpression of ssDNA-annealing proteins (RecT-like) naturally found in the genomes of the above organisms or their close relatives. The efficiency of the ssDNA oligo-driven recombination can be increased by selective removal of cellular functions associated with mismatch repair.

Multiple applications of recombineering approaches have further expanded science capabilities. For example, the *Lambda Red* system was used to generate the *Keio* collection of *E. coli* knockout mutations of nonessential genes, where each gene is replaced with a cassette coding for kanamycin resistance gene. The collection has enabled many different *E. coli* genomic studies such as creating maps of genetic interactions and identifying sets of genes instrumental for the bacterial response to various stress conditions and drug treatments. Recombineering in *E. coli* has enabled development of experimental strategies for accelerated evolution, genome-wide codon replacement, designing of minimal genomes, all of which are expected to further propel synthetic biology. Metabolic engineering is another area that has been transformed by recombineering techniques, which have allowed for straightforward and efficient removal of genes inhibiting metabolite production, modulation of expression of certain enzymes by altering the promoters and regulatory elements of the genes coding for them, or insertion of entire biosynthetic pathways in new hosts. Developing protocols for recombineering in pathogenic and fermenting bacteria is expected to result in engineering of vaccine strains and high-performing industrial strains, respectively. *Lambda Red* system has contributed to the advancement of mouse genomics and genetics by enabling fast and precise modifications of eukaryotic DNA sequences cloned in BACs. Together with the applications of the site-specific recombinase Cre from bacteriophage *P1* and the increasing collaborative efforts in the field, science now has the capabilities of generating transgenic mice and mouse models of disease in much faster and more cost-effective ways. The addition of gene editing approaches in the toolbox of mouse genome manipulations creates unprecedented opportunities for advancement of many areas of mammalian molecular biology.

3.2.3 Cre/LoxP Recombination System

The natural function of the Cre/LoxP recombination system although essential for the biology of bacteriophage *P1* is rather "boring" in comparison to its impact to research advancements, especially in the field of generating transgenic mice. *P1* establishes lysogeny without integration in the host chromosome. Instead, it persists in the cytoplasm as a single copy circular episome, which is duplicated in a coordinated manner with cellular division. The *P1* genome duplication produces two interlocked dsDNA circles, which are resolved by the Cre recombinase. Cre promotes recombination reaction between the two LoxP sites positioned as direct repeats in the phage genome. The LoxP site consists of only 34 nucleotides, which can be further broken down to left and right inverted repeats, 13 nt each, and an 8-nt-long spacer separating them. The Cre molecules bind to the repeats of the recombining LoxP sites and catalyze a crossover within the spacer sequence. Complete spacer homology is a requirement for efficient reaction. Recombination can take place between LoxP sites positioned on the same molecule or on different molecules. Cre does not require any accessory factors, and it has been shown to be active in many heterologous systems including mammals and plants. Depending on the number and the orientation of the LoxP sites, Cre-mediated recombination can generate insertions/deletions, inversions, translocations, and sequence (cassette) replacements (Fig. 3.6). The Cre/LoxP system is most frequently used to delete genes of interest, i.e., creating gene knockout mutants or removing selectable markers, or to insert genes of interest, i.e., creating knock-in mutants or sequence replacements. Sequences flanked by LoxP sites are described with the term "floxed."

Cre-driven deletions are most frequently utilized to create conditional mutants in transgenic mice or to remove selectable markers from transgenic plants. The simplest approach involves a complex scheme of molecular biology manipulations and classical breeding techniques. Initially, a transgenic mouse coding Cre under the control of tissue-specific promoter is created using conventional approaches. In a separate experiment, a mouse coding for a floxed version of the gene to be removed is created. The Cre reaction takes place after the sequence coding the enzyme and the floxed gene are brought together by conventional breeding of the two transgenic mice. Cre expression is induced, and the enzyme excises the targeted gene creating a deletion mutant that can be used to address various research questions relevant to the function of the deleted gene (Fig. 3.7). Alternatively, the Cre coding sequence can be delivered to a mouse carrying the floxed gene of interest (GOI) by viral vector. The introduction

Inversion

Deletion/Insertion

Translocation

Cassette exchange

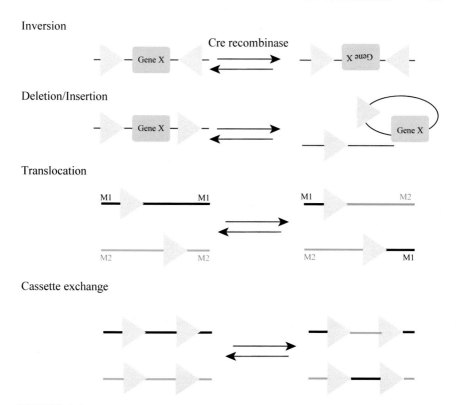

FIGURE 3.6   **Recombination reactions supported by the Cre/LoxP recombinase of bacteriophage P1.** The Cre/LoxP system catalyzes reversible site-specific recombination reactions between sequences flanked by LoxP sites (depicted with *gray triangles*). Depending on the number and the orientation of LoxP sites Cre-driven reactions can result in inversions, deletions, insertions, translocations, or cassette exchanges.

of the Cre/LoxP technology in the protocols for generation of transgenic mice significantly streamlined the process. In addition, multiple international consortia exist, sharing Cre mouse strain resources, data, and experience, focused on making the process more efficient and creating collections of transgenic mice with relevance to particular diseases, organs, etc. Introduction of clustered regularly interspersed palindromic repeats/ CRISPR-associated enzyme 9 (CRISPR/Cas9) technology is a game changer for the transgenic mice field reducing the time for strain generation to several months. Side by side, comparisons of transgenic mice models of sarcoma generated with Cre/LoxP technology and CRISPR/Cas9 technology have demonstrated that induced sarcomas are very similar.

Conceptually, the same strategy is used to remove selectable markers used in the *Ti* plasmid–driven generation of transgenic plants. By design the gene for the introduced selective marker is floxed. When a

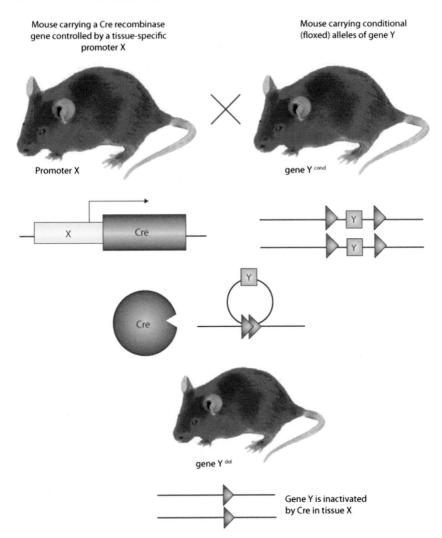

FIGURE 3.7 **Conceptual application of Cre/LoxP recombination system to create conditional mutations in transgenic mice.** Transgenic mouse capable of expressing Cre from inducible promoter and a mouse coding for a floxed GOI are created using conventional approaches to generation of transgenic animals. Both mouse strains are mated and Cre expression induced in a developmental point of interest, resulting in a conditional null mutant. *Reprinted from Rosenthal N, Brown S. The mouse ascending: perspectives for human-disease models.* Nat Cell Biol 2007;9(9):993–9, *with permission from Nature publishing group.*

transgenic plant coding for GOI and floxed selective marker is crossed with a plant coding for Cre under the control of chemically inducible promoter, the marker can be eliminated in the F1 progeny when the hybrid plants are treated with a substance inducing Cre expression. An alternative strategy is autoexcision, which shortens the experimental procedure of marker removal. In this scenario, Cre is coded on the same DNA fragment as the selective marker, and both coding sequences are floxed. Cre is cloned under the control of tissue-specific or chemically/heat inducible promoter, which when turned on triggers enzyme expression and subsequent removal of the marker gene along with the Cre coding sequence. The autoexcision approach is the only one applicable to plants propagated vegetatively (asexually) in commercial settings. Cre-driven deletions can also be introduced in the context of a cell culture, where constructs containing relevant sequences are delivered in the cell by transfection of plasmids or transduction with viral vectors.

Cre-mediated transgene integration follows similar key steps. In its simplest form, the approach calls for generation of cell lines or organisms harboring LoxP site using conventional methods, thus creating a LoxP target site in a defined location, which permits precise and controlled insertion. A promoter of choice is engineered upstream of the LoxP site. When a promoterless DNA fragment harboring a LoxP site, a selection marker, and a GOI is introduced in the system, recombination takes place and the heterologous genetic material is expressed. An alternative strategy is the recombination-mediated cassette exchange, in which cells and organisms harboring a cassette flanked by LoxP sites is created as a "target cassette." Recombination reaction exchanges the DNA sequence positioned between the two LoxP sites, effectively introducing a GOI in the targeted location. Cre recombination is reversible, which presents a challenge for genome manipulations aiming at stable insertion of fragments. The issue is circumvented by expressing Cre only transiently and/or using pairs of wild-type and truncated LoxP sites or pairs of LoxP sites harboring spacer mutations. Cre/LoxP recombination is highly specific, although some level of nonspecific reactions have been reported with variable frequency in different model systems and cell types, most likely in correlation with existing pseudo-LoxP sites. Considerable efforts have been invested in tailoring Cre properties for recombineering purposes. For example, Cre coding sequence has been changed to make it more eukaryotic-like and to reduce CpG content. Fusion with estrogen receptor ligand-binding domain and other proteins has been engineered to allow more controlled expression; Cre gene has been engineered in the context of various inducible promoters with different regulatory elements to ensure optimal expression in a broad range of model system. In vitro approaches for recombination-based cloning utilizing Cre/LoxP have been also developed.

3.2.4 Phage Integrases

Bacteriophage integrases are site-specific recombinases whose natural purpose is to insert and excise the viral genome during the establishment of lysogeny and the transition from lysogenic to lytic life cycle.

The integration process is highly specific and is executed solely by the activity of the integrase enzyme. In contrast, the excision employs additional proteins known as excisionases (Xis) in model systems utilizing tyrosine integrases and recombination directionality factors in model systems utilizing serine integrase. The designation of recombinases as tyrosine or serine type is based on the nature of the key amino acid for catalysis. Among the best studied integrases is the one from bacteriophage lambda, a member of the tyrosine family, whose overall mode of action is depicted on Fig. 3.8. The enzyme binds to the two recombination substrates *attB*, found in the *E. coli* genome, and *attP*, found in the phage genome and brings them together. DNA cleavage and strand exchange follow resulting in Holliday junction intermediate (not shown in the figure), which is resolved to form a recombinant molecule that can be described as an insertion of the phage genome into the bacterial chromosome. The phage genome is flanked by two recombinant sites, each containing half of *attB* and *attP* recombination substrates. The site on the left of the inserted phage is designated as *attL*, whereas, the one on the right as *attR*. A cellular protein, IHF (integration host factor), facilitates recombination by bending DNA and thus bringing the participating DNA strands in close proximity. The excision reaction takes place via similar steps and requires two additional accessory factors: Xis encoded by the phage and Fis, encoded by the host. Int, IHF, Xis, and Fis form a complex, which specifically binds to the P region of *attR* and promotes DNA cleavage and strand exchange recovering the original *attB* and *attP* sites, thus effectively executing clean and scarless removal of the

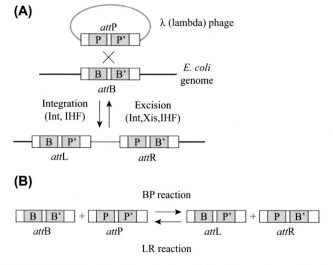

FIGURE 3.8   Mechanism of bacteriophage lambda integration and excision.

phage. The properties of integrases have been applied toward development of approaches for chromosomal integration of transgenes in *E. coli*, *Staphylococcus aureus*, *Pseudomonas* (among others); in vitro universal cloning (the *Gateway* system) and clonetegration techniques combining cloning and chromosomal integration in one approach. Increasing number of phage integrases are being utilized offering different options to researchers. For example, in addition, to *lambda* integrase enzymes from bacteriophages P21, HK022, 186 (all tyrosine integrases) are available as options for DNA integration based on the natural *attB* sites of these viruses. The *attB* site of the *Streptomyces PhiC31* bacteriophage has been engineered in *E. coli* allowing a heterologous integrase to be utilized. The latter is a serine recombinase, which offers the added advantage of short *att* sites (less than 50 nt, in contrast to ~200 nts *attP* sites for tyrosine integrases) and no need of host proteins such as IHF in the *lambda* system. The PhiC31 integrase has been used successfully to modify the genome not only of heterologous prokaryotic species but also to engineer a growing list of eukaryotic genomes including mammalian (mouse, hamster), insect (silkworm, *Drosophila, Anopheles*), and plant (*Arabidopsis*, wheat) genomes. It has been shown that another serine integrase isolated from the *Mycobacterium smegmatis* phage Bxb1 is also compatible with multiple eukaryotic environments. A combination of Cre/LoxP and the phiC31 integrase has been utilized in tandem to introduce GOIs and selective markers (Cre/LoxP) and subsequently to remove the marker and the LoxP site (phiC31 integrase), resulting in irreversible insertion of GOI with minimal addition of DNA sequences. For historic reasons, the lambda integrase continues to be the most used integrase, although it is expected that more and more serine integrases will find their spot in the landscape of molecular biology.

The *Gateway* system for universal cloning utilizes the lambda integrase biology and a broad pool of vector variants allowing for fast, efficient, and easy transfer of GOI initially cloned into an entry vector to multiple destination vectors with different properties supporting *E. coli*, mammalian or insect expression, tagging, etc. The latest addition to the system (as of 2017), *Multisite Gateway*, is a variant allowing simultaneous cloning of up to three genes on one vector in order of interest. The system takes advantage of the BP (insertion) and LR (excision) recombination reactions (Fig. 3.8) driven by the corresponding recombination sequences and clonase enzyme mixes, as well as the *ccdB* marker (Fig. 3.9). The *ccdB* gene codes for toxin from one of the *E. coli* toxins–antitoxin systems. In the absence of its antitoxin coded by ccdA, it kills the cell by interfering with the function of gyrase/topoisomerase II, an essential bacterial enzyme. By design the expression of CcdB is possible only from constructs carrying by-products of the recombination reactions and thus the cells harboring them are eliminated from the experimental flow. *Gateway* cloning

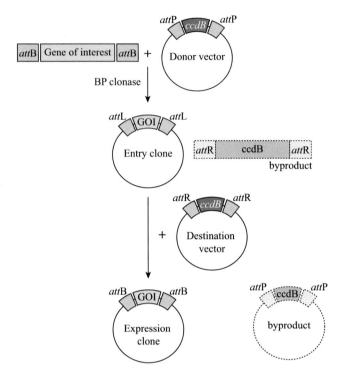

FIGURE 3.9 **Conceptual application of lambda integration machinery in Gateway cloning.** The gene of interest (GOI) is amplified by PCR using primers containing *attB* sequences. The resulting DNA fragment is recombined with a donor vector coding for the *ccdB* selective marker flanked by *attP* sequences in a reaction driven by BP Clonase mix catalyzing a lambda insertion type reaction (BP reaction). The product of the BP reaction is an entry clone in which the GOI is flanked with *attL* sequences. The entry clone can be recombined with an array of destination vectors, each coding for specific target feature (for example, expression in insect cells or His-tag fusion) and *ccdB* marker flanked by *attR* sequences. A recombination reaction driven by LR clonase mix, an LR/excision reaction, yields a clone of interest. Cells carrying by-products (depicted in fading gray) of the recombination reaction are eliminated by the toxic action of the *ccdB* marker. Unidirectionality of the cloning is achieved by using slightly different versions of each type *att* site on the left and the right side of the GOI.

proceeds in two stages: (1) generation of entry clone coding for a GOI by BP recombination between the *attB* sites engineered at the ends of the PCR product to be cloned and the *attP* sites flanking the CcdB marker on the donor vector and (2) LR reaction to recombine the entry clone with a destination vector of interest resulting in the final product (depicted as an expression clone in the example shown on Fig. 3.9). Both steps employ the CcdB marker, which is eliminated from the system by the virtue of its own lethal action. The Gateway technology is fast and efficient and allows the same entry clone to be recombined with multiple destination vectors yielding final constructs with various properties.

3.3 GENOME EDITING

3.3.1 Genome Editing Overview

Genome editing is the newest area of genetic engineering, which provides an excellent example for the transformative power of science and the fast pace of moving basic science discoveries in the realm of practical applications. In 2011, *Nature Publishing Group* selected gene editing as method of the year, whereas in 2015 the journal *Science* declared the CRISPR/Cas9 system a breakthrough of the year. Concurrently, concerns for proper use of the new technology are being voiced from the point of view of bioethics, potential use to engineer pathogens as weapons for mass destruction and release of genetically modified organisms in the environment. Undoubtedly, gene editing has the potential to test the limits of current legislation and beliefs on the safety and ethics of biomedical interventions.

Mechanistically, genome editing is a collection of approaches that enable fast and precise insertion, deletion, and replacement of genetic material in vivo by introducing sequence-specific double-strand breaks at genome sites of interest and guiding their repair through nonhomologous end joining (NHEJ) or HR generating specific genome edits/mutations (Fig. 3.10). NHEJ pathway repairs dsDNA breaks by joining both ends directly, which most frequently results in small deletions/insertions designated as indels. Homologous repair pathway uses donor template to repair dsDNA breaks and results in full recovery of the molecule, assuming the repair template is identical with the original molecule. Variations in the nature of the repair template translate in insertions, deletions, or point mutations and present the desired outcome from gene editing. Four groups of site-specific nucleases can be employed to introduce sequence-specific dsDNA breaks: meganucleases, zinc-finger nucleases (ZFNs), transcription activator–like endonucleases (TALENs), and the CRISPR/Cas9 system. This subsection will focus mainly on the properties and the applications of the CRISPR/Cas9 system and will only briefly discuss the other three.

3.3.1.1 Meganucleases

Meganucleases, also referred to as homing endonucleases, are found in bacteria, phages, archaea, and some eukaryotes and are often coded by introns or inteins that have the capability to act as mobile genetic elements. The designation meganucleases originate from the rather large size of their DNA recognition sequence ranging between 20 and 45 nucleotides. Meganucleases target sites tend to be found in genomic regions lacking introns and inteins. The enzymes introduce dsDNA cuts, which are then repaired using HR and the intron/intein coding sequence as

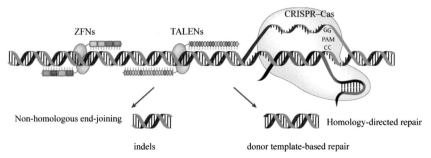

FIGURE 3.10   **Overview of nuclease-driven gene editing.** Nuclease-driven gene editing approaches work by targeting sequences of interest with sequence-specific nucleolytic activity delivered by zinc-finger nucleases (ZFNs), TALENs, or CRISPR/Cas9, which introduce controlled double-stranded DNA breaks. Meganucleases (not shown) can also be used for the same purpose, however, with very limited utility. The breaks are repaired by the cell using the natural repair mechanisms of the cell: nonhomologous end-joining mechanism, resulting in indels (on the **left**) and homology-directed repair mechanism (on the **right**), which can result in sequence recovery or various types of edits (mutations) depending on the sequence of the donor template. PAM stands for protospacer adjacent motif with the consensus sequence 5′NGG3′, from which only the last two nucleotides are depicted. *Reprinted from Haimovich AD, Muir P, Isaacs FJ. Genomes by design.* Nat Rev Genet *2015;16(9);501–16, with permission from Nature publishing group.*

template, thus introducing a new copy of the intron/intein at the targeted site. Meganuclease-mediated gene editing depends on the presence of the corresponding DNA recognition site in the area of the genome where editing is needed, which is a serious limiting factor. Currently several hundred meganucleases are known, and methods for evolving their specificity have been described, thus potential for developing a broad array of meganucleases exists. However, the advancements and simplicity of the CRISPR/Cas9 system offers tremendous advantages for fast and technically easier editing, making meganucleases a tool with rather low utility. Biologically, phage meganucleases are thought to give competitive advantage to the phages that code them in mixed infections by cutting the genomic DNA of the coinfecting phages. Interestingly, bacteriophage T4 codes for 15 different meganucleases, which take approximately 11% of the genome coding capacity.

3.3.1.2 Zinc-Finger Nucleases

ZFNs and TALENs are chimeric molecules comprised of the nuclease domain of the FokI restriction enzyme and site-specific DNA-binding domain based on zinc-finger motifs (in ZFNs) or the transcription activator–like effector proteins (in TALENs; Fig. 3.10). FokI is a dimeric-type IIS restriction enzyme isolated from *Flavobacterium okeanokoites*, which recognizes the 5′-GGATG-3′ sequence and introduces two single cuts 9 nt away from the 3′ end of its recognition sequence on the top strand (listed above) and 13 nt away from the 5′ end of the bottom strand sequence (complementary to the listed one), thus collectively creating sticky ends with 4 nt-long overhang. The mechanism of its action allows restriction to take place essentially in any sequence, as long as it is positioned at the precise distance from the recognition site bound by the DNA-binding domain. ZFNs and TALENs are essentially engineered restriction enzymes employing the nuclease domain of FokI and zinc fingers or TAL effector DNA-binding domains to tether the nuclease activity at a sequence of interest. Huge advantage of ZFN and TALEN approaches is the relatively straightforward way to manipulate their specificity, thus theoretically making it possible to direct editing efforts to any sequence.

Zinc fingers are small structural protein motifs that coordinate zinc, which are commonly found in proteins taking part in nucleic acid–based processes in the cell. Initially, zinc-finger motifs were attributed to transcription factors; however, later they were found in many proteins interacting with DNA/RNA-binding proteins. It is thought that their principle function is to confer specificity of DNA binding directly (in transcription factors) or indirectly by binding to DNA/RNA-binding proteins and altering their specificity. DNA specificity of zinc fingers can be engineered or selected for in phage display screens. ZFN molecules are engineered to contain a tandem of 3–6 zinc-finger motifs collectively recognizing a sequence of 9–18 nucleotides (each motif accounts for 3 nucleotides). Engineering a pair of tandem repeats specifically binding to the top and bottom target DNA strand allows the attached FokI nuclease domains to be placed in a position to introduce cuts in target of interest. A limitation of ZFNs is the potential for overlapping specificities of the employed zinc-finger motifs, which can lead to off-target cutting. Tools for design and selection of motifs with the desired specificity have been developed, and ZFNs have been successfully used for proof of principle editing in many model systems, as well as to solve problems with therapeutic significance. For example, immune cells harboring *ccr5* deletion mutation have been engineered, thus rendering them resistant to HIV infection. Clinical trials based on ZFN-driven genome corrections are currently in progress.

3.3.1.3 *Transcription Activator–Like Endonucleases*

TAL effectors are eukaryotic transcription factors secreted by bacterial pathogens infecting plant cells. They penetrate the cell, enter the nucleus, and selectively activate expression of individual genes. The best studied TAL effectors are isolated from *Xanthomonas* bacteria, which are rather common pathogen affecting multiple plant species and causing serious economic losses. Recently TAL effectors have been described in *Ralstonia solanacearum*, a key plant pathogen commonly used as a model system to study host–pathogen interactions and have been designated "RipTALs."

TAL effectors are characterized with a modular architecture that includes a DNA-binding region consisting of tandem repeats (33–34 residues each), which are nearly identical with the exception of the position 12 and 13 designated as repeat variable diresidues or RVDs. Each RVD specifies the identity of a single base thus making the tandem array of repeats high-specificity determinant for the DNA binding of the molecule as a whole. The "code-like" relationship between amino acid sequence of RVD and recognized/bound nucleotides allows for engineering of specific DNA-binding domains in a manner very similar to the assembly of structures with Lego bricks. DNA-binding domains with desired specificity can be assembled by selecting a combination of repeats with particular RVDs. Interestingly, the RVD code in *Xanthomonas* and *Ralstonia* species is slightly different. The modular structure of TAL effectors allows different functional parts of the original molecules to be seamlessly exchanged with domains and sequences with completely different functionalities. For example, replacement of the TAL effector transactivating domain with the nuclease domain of the FokI restriction enzyme allows TALENs to be engineered. Replacement of the plant nuclear localization signal with heterologous one enables TALENs to be used in animal cells. Compared with meganucleases, ZFNs and TALENs offer significantly larger potential for easy specificity retargeting. Both ZFNs and TALENs require expression of two individual protein variants (one specifically binding to each DNA strand) in order productive dsDNA break to be generated. Nickase versions of ZFNs and TALENs introducing a cut only in one of the DNA strands of dsDNA have been engineered and have been found to direct the subsequent repair mostly toward HR pathway, which is a highly desirable outcome; however, the frequency of editing events in nickase-based experiments is lower compared with experiments with the variants introducing dsDNA breaks. Ex vivo editing with TALENs and subsequent cell transplantation has shown promise in experimental treatment of leukemia. Safety and efficacy clinical trials of TALEN-based protocols are currently underway.

3.3.1.4 The CRISPR/Cas9 system

The CRISPR/Cas9 gene editing system is derived from prokaryotic cells "immunity system" managing adaptive response against viral infections (see Section 3.3.3.1). Its unprecedented utility in gene editing and the boom of relevant applications have overshadowed the appreciation and significance of its natural purpose only a few years after its discovery. The huge promise for straightforward and efficient editing comes from the simplicity of the technology and the ease of technical manipulations in comparison with any other editing approach. The CRISPR/Cas9 system employs two key molecules: a guide RNA (gRNA) and an RNA-guided nuclease Cas9, which introduces blunt dsDNA break in the target sequence 3 base pairs upstream of the protospacer adjacent motif

(PAM, see below), which is composed of the 5′NGG3′ sequence, found quite frequently throughout genomes. The huge advantage of CRISPR/Cas9 comes from the straightforward way specificity is achieved: simply by conventional base pairing of the gRNA to the target DNA region, thus allowing editing to take place virtually anywhere in the genome as long as gRNA is designed to base-pair properly in respect to the PAM sequence. Unlike ZFNs and TALENs, there is no need to design specific zinc finger or tandem repeats for each editing target and attach them to FokI nuclease domain. One can simply design a gRNA for each target, a process very similar to designing primers for PCR, and repeatedly utilize a molecular tool delivering a functional Cas9 nuclease in the cells subject to genome editing. First clinical trials involving CRISPR/Cas9 editing were approved in 2016.

Side-by-side comparisons of ZFNs, TALENs, and CRISPR/Cas9 gene editing are starting to emerge and contribute to the better understanding of their advantages and limitations. Theoretically, the same principle outcomes can be achieved with the three technologies, and their efficiencies have been found to be generally similar. At the same time, emerging data demonstrate that efficiency and precision are quite variable across organisms, cell types, and procedures, calling for further examination and caution when extrapolating findings observed in one system to expected outcomes in another system. Biologically, the CRISPR/Cas9 system has the advantage that it is insensitive to methylation and thus can edit both methylated and unmethylated DNA, which is an important feature for applications in eukaryotic cells. CRISPR/Cas9 is also multiplexing-friendly approach allowing multiple edits to take place simultaneously, assuming multiple guiding RNAs are provided at the same time. Technologically, the CRISPR/Cas9 system has significant advantages due to its simplicity, versatility, and cost-effectiveness. These advantages are stemming from the lack of protein engineering step, as well as the straightforward way to specifically target sequences of interest. The versatility and the cost-effectiveness are in part driven from the open access policy of the CRISPR research community and open sharing of resources and experience that promote collaborative efforts to better understand the system and develop more practical applications. Fig. 3.11 here provides a snapshot of the impact of nuclease-based editing technologies on research as judged by number of publications and resources requests/depositions. Despite the will of sharing, patent wars are still in progress (as of 2017) to settle the intellectual property rights of different aspects of the CRISPR/Cas technology and will likely influence how scientist approach new technology development in the future.

A major issue with gene editing technologies is steering outcomes away from the NHEJ pathway of repairing the dsDNA cuts introduced in the targeted genome. NHEJ repair results in indel mutations that can inflict

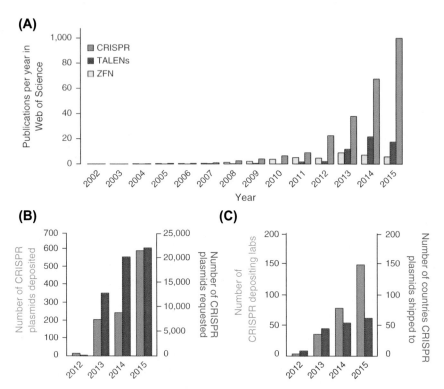

FIGURE 3.11 Impact of gene editing. (A) Number of manuscripts listing CRISPR, ZFN, and TALEN as keywords published between 2002 and 2015 according to Thomson Reuters' Web of Science. (B) The number of CRISPR constructs deposited and requested at Addgene, a nonprofit global plasmid repository based in Cambridge, MA. (C) The number of Addgene-depositing laboratories and recipient countries. *Reprinted from Barrangou R, Doudna JA. Applications of CRISPR technologies in research and beyond.* Nat Biotech *2016;34(9):933–41, with permission from Nature publishing group.*

disruption of critical genes if accidental off-site cuts take place. Cuts in the intended target sequences are repaired by the HR pathway when a donor repair template is provided together with the guiding RNA and Cas9 nuclease. Unpredictable off-site cuts influence different applications of CRISPR/Cas9 editing system to different extent. For example, editing in bacteria can be monitored in a much more straightforward way than in eukaryotic cells. Editing of eukaryotic cells ex vivo offers much better options for safety assurances before the edited cells are transplanted in a patient in comparison with in vivo editing performed directly in the patient's body. Similarly, gene editing of germ-line cells and embryos poses different level of requirements for control of accidental off-site DNA cleavages. Correspondingly, reducing the rate of off-site cleavages is an active area of research for all editing technologies, especially

CRISPR/Cas9, which was most recently developed and thus the impact of the issue is still to be fully appreciated. Development of nickase versions of the editing nucleases has been shown to lower off-site cleavages, as well as the optimization of nuclease level of expression. Experiments varying the length and the sequence composition of the gRNA are being performed in attempt to identify possible rules/biases in the biochemistry of the target DNA/gRNA recognition, annealing, and cleavage that can be utilized in a future rational design approaches. Parallel efforts are focusing on development of diagnostic assays for off-site targeting, allowing straightforward testing before gene editing applications. Another common issue across the board of the gene editing technologies is the mode of delivering of the gene editing tools to different types of cells and/or organisms. While the process is relatively straightforward for prokaryotic systems, especially the ones established as traditional model systems, many constrains exist in eukaryotes. Regardless of existing challenges, most likely no one doubts that gene editing is here to stay. How far the range of its applications will extend is a function of finding satisfactory solutions to the above challenges, changes in policy and social perceptions. Beyond bioethics concerns regarding biomedical applications, it is unclear if edited plant with economic significance to food supply would be classified as genetically modified organism and subjected to current regulations or will receive a different designation and an exempt status. It is also not clear if it is feasible edited organisms from gene drives (see below) to be freely released in the environment as a tool to rebalance populations. It is quite likely gene editing to become one of the pieces of technology in human history for which the future arrives in flying colors before the society is ready to fully appreciate the magnitude of its impact.

3.3.2 Key Applications of Nuclease-Based Editing

Mechanistically speaking, two key characteristics of the nuclease-based editing technologies are essential for their practical applications: (1) sequence-specific nuclease activity and (2) sequence-specific DNA binding. Each one of them has been used for the development of array of practical approaches with applications in diverse fields driven from the natural importance of altering DNA sequences or being able to tether molecules with particular properties at specific sites. Depending on the sequence of the donor template provided for the HR-based repair of the introduced dsDNA breaks, gene editing allows engineering of deletions, insertions, introduction of point mutations, and corrections of point mutations. These strategies have been used successfully to create countless number of research reagents; to alter or remove genes conferring sensitivity to pathogens; to correct genetic defects resulting in disease and create corrective gene therapy tools; to stack up beneficial genes in plants allowing them to be inherited together in classical breeding approaches; to engineer key metabolic pathways in commercially relevant organisms; to create disease models on the cellular and organismal level, among many other examples. A separate group of applications takes advantage of the ability of ZFNs, TALENs, and CRISPR/Cas9 system to specifically bind to DNA. For example, transcription activators and repressors have been designed by utilizing the sequence-specific DNA-binding determinants of the gene editing nucleases and targeting them to promoters or regulatory elements of interest (Fig. 3.12B and C, the diagrams are drawn on the example of CRISPR/Cas9 system). In the case of CRISPR/Cas9, this is accomplished by using a Cas9 mutant with abolished nuclease activity. Different protein entities can be fused to the mutant Cas9 protein and targeted to sequences specified by the chemical nature of the gRNA. For example, transcription activator proteins can influence transcription rates of targeted promoter/gene, methylases can alter the methylation status of specific chromosome regions; fluorescent proteins can be used to image the location of corresponding DNA sequences. A third group of applications is based on the selective elimination of particular genotypes in vivo (Fig. 3.12D–E). For example, pathogenic bacterial strains can be selectively eliminated by the CRISPR/Cas9-based bacteriocide, which results in cleavage of genetic material associated with pathogenicity such as bacterial toxins; recombinant genotypes can be favored by CRISPR/Cas9 targeting of wild-type ones; and gene drives can be created to influence the prevalence of particular gene alleles in a population. The approach has been applied toward elimination of genes associated with herbicide resistance, altering the reproductive capabilities of invasive species or insect disease vectors.

3.3.3 Biology of the CRISPR/Cas-Based Adaptive Immune System in Prokaryotes: The Big Picture

Despite its popularity and utility as a gene editing system, CRISPR/Cas natural biological function has little to do with genome editing the way it is taking place in research laboratories. CRISPR/Cas is an elaborate and dynamic molecular system that allows bacteria and archaea to protect themselves from invading viruses, plasmids, and transposons by enzymatically cleaving their nucleic acid. Molecular record of the encountered invaders is kept by acquiring short pieces of their genetic material and inventorying it in a specific format in the CRISPR locus. The molecular record sequence is transcribed into a gRNA, which provides a template for nucleolytic cleavage of the same invading nucleic acid if the corresponding virus/plasmid/transposon revisits the cell again. Currently, CRISPR/Cas systems are thought to exist in approximately 40% of bacteria and 90% of archaea as appreciated by bioinformatics approaches. Characterization of CRISPR systems from different species has revealed

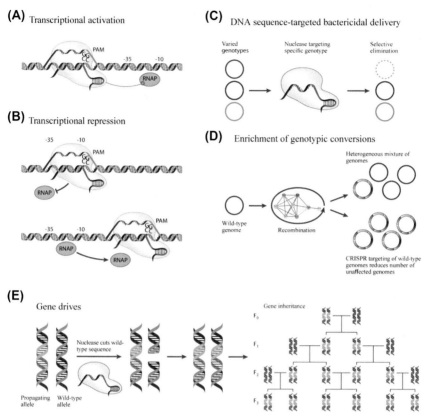

FIGURE 3.12 Key applications of gene editing. Gene editing technologies have been applied toward many different research scenarios and tool development. Key concepts are depicted on the example of CRISPR/Cas9 editing; however, they are technically possible with all editing approaches. Transcription activation (A) and repression (B) tools are based on nuclease-deficient Cas9, which is able to support specific DNA binding without cleaving DNA. Activation is achieved by engineering Cas9 to participate in protein–protein interactions with RNAPol (RNAP) effectively recruiting it to promoter of interest. Transcription repression is achieved by designing the gRNA to base-pair with promoter regions (inhibiting initiation) or with sequences immediately after the promoter (inhibiting elongation), thus allowing the transcription of group of genes or individual gene to be inhibited. Selective elimination of alleles of choice is used as a strategy to eliminate pathogenic bacterial strains, while preserving nonpathogenic ones (C); to enrich mixtures with recombinant bacterial genomes while selectively eliminating nonrecombinant genomes (D); or to selectively propagate alleles of interest in gene drives (E). *Reprinted from Haimovich AD, Muir P, Isaacs FJ. Genomes by design. Nat Rev Genet 2015;**16**(9):501–16, with permission from Nature publishing group.*

tremendous diversity of components and mechanisms, which continue to be studied extensively and the obtained knowledge to be applied to innovative research approaches and technologies. The majority of described CRISPR systems target invading dsDNA, although systems able to cleave invading RNA have been reported.

Currently, CRISPR systems are classified into two major classes and six major types with several subtypes based on the machinery employed. Class I CRISPR/Cas systems (encompassing types I, III, and IV) employ multicomponent complexes to execute the nucleolytic cleavage of invading nucleic acid, whereas Class II systems (encompassing types II, V, and VI) employ a single nuclease to do so. The CRISPR/Cas9-based technology for gene editing belongs to Class II/type II. Its fast development and wide application is largely due to its simplicity and the minimal number of employed components. The description of CRISPR/Cas-based immunity will focus on the big picture and CRISPR/Cas9 as opposed to rigorous presentation of all variants and details of existing CRISPR systems.

3.3.3.1 CRISPR-Based Adaptive Immunity: Components and Mechanism

The CRISPR/Cas systems (Fig. 3.13) encompasses an array of CRISPR spacers (depicted with color boxes numbered 1–6) and repeats (depicted with white boxes) flanked by genes coding for Cas proteins (depicted with blue arrows). Each spacer is a short (30–45 nt) sequence homologous to foreign DNA that has invaded the cell previously, i.e., each spacer can be viewed as a snapshot of an intruder "taken" from the surveillance camera of the bacterial immune system. The original sequence in the phage genome corresponding to the acquired spacer is designated as a protospacer. Each CRISPR array starts with a leader DNA sequence (a gray box designated with letter L) guiding the integration of new spacers.

The action of CRISPR/Cas adaptive immunity is executed in three stages: (1) spacer acquisition, (2) transcription, and (3) interference. Spacer acquisition could be considered as an equivalent of immunization in the human immune system equipping the cell to defend itself against a repeated invasion. The process of protospacer identification is best understood in CRISPR/Cas type I systems, although many details are still lacking. It is believed that fragments of phage DNA are generated by double-stranded DNA breaks randomly arising in viral replication. Protospacer sequences are presumably identified by Cas enzymes based on recognition of 2-nucleotide-long to 3-nucleotide-long motifs, known as PAMs. The spacer integration in the CRISPR array is accomplished by two coordinated cleavage/ligation reactions taking place at the 5′ end of the first repeat sequence downstream from the leader. The spacer is ligated to the cleavage-generated 5′ ends resulting in its incorporation in the CRISPR array immediately after the leader sequence in the context of two ssDNA gaps corresponding to the flanking repeats. The gaps are repaired by DNA polymerase and sealed by a ligase, which complete the spacer integration process. Cas1 and Cas2 nucleases play key role in the

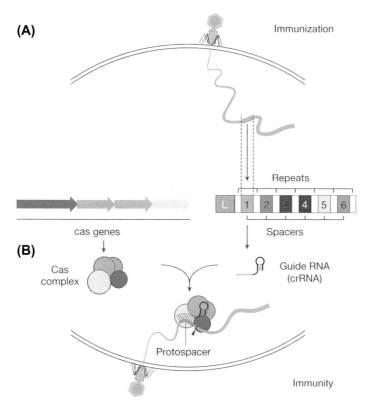

FIGURE 3.13 **Mechanism of CRISPR/Cas adaptive immunity.** CRISPR loci consist of short spacer sequences of phage and plasmid origin (colored numbered *boxes*) separated by DNA repeats (*white boxes*) arranged into an array beginning with a leader sequence (*gray box* labeled with the letter L) containing the promoter for its expression. The array is flanked by a Cas operon (*blue arrows*) coding for the associated protein components, which vary by number and characteristics depending on the type of the immunity system. (A) Immunization step: spacer sequences are captured as they enter the cell and integrated into the CRISPR array immediately after the leader sequence. (B) Immunity in action: the CRISPR array is transcribed in a long transcript, which is subsequently processed into individual CRISPR RNAs (crRNAs), each coding for one spacer. crRNAs form complexes with Cas RNA-guided nucleases and target them to complementary protospacer of invading bacteriophage, which is subsequently cleaved. *Reprinted from Marraffini LA, CRISPR-Cas immunity in prokaryotes. Nature 2015;526(7571):55–61, with permission from Nature publishing group.*

spacer integration process and are well conserved among most CRISPR/Cas systems described up-to-date.

The CRISPR array locus is transcribed into long precursor RNA molecule, which is processed to individual crRNA (CRISPR RNA) molecules, each containing one spacer sequence. Mature crRNAs are able to hybridize to complementary protospacers when relevant bacteriophage DNA enters the cell, thus serving as a guide for the subsequent nucleolytic cleavage and blocking the phage infection by crRNA-based interference. One can think about the transcription stage of the CRISPR/Cas immune system as a process of preparing tools capable of neutralizing future phage infections, i.e., a process conceptually resembling B lymphocyte maturation leading to the synthesis of antigen-specific neutralizing antibodies. Similarly, the interference stage conceptually resembles the pathogen elimination by neutralizing antibodies. Interference takes place by different mechanisms in different CRISPR/Cas systems. The CRISPR/Cas9 system employs a single protein, Cas9, and a transactivating RNA molecule, commonly described as tracrRNA, i.e., transactivator of crRNA. The 5′ end of the tracrRNA is complementary to the 3′ end of the crRNA, thus their base pairing results in short dsRNA segment. It is believed that tracrRNA contributes to crRNA maturation. If the sequences of the crRNA and tracrRNA are fused together the resulting RNA piece, commonly described as gRNA is completely functional. That discovery allowed for the very straightforward design of the CRISPR/Cas9 gene editing tool, which essentially requires one protein component, Cas9 (utilized in all editing reactions), one RNA component (gRNA) specific for each sequence subject to editing and one donor DNA template to be used to repair the CRISPR/Cas9 introduced dsDNA break. The simplicity of CRISPR/Cas9 system and the availability of established approaches to deliver the two needed components into target cells and multicellular organisms allowed for the fast paced advancement of CRISPR/Cas9-based applications.

Cas9 is a multidomain protein harboring PAM-recognition domain, two individual nuclease active sites along with gRNA- and target DNA-binding activities. The protein by itself is inactive until it forms a complex with the gRNA, which allows it to start scanning or "interrogating" DNA. Nuclease cleavage does not take place unless the Cas9/gRNA complex recognizes and base pairs with a protospacer and PAM sequence simultaneously. The base pairing does not need to be perfect. The PAM consensus sequence is NGG (less frequently NAG) and the spacer/protospacer base pairing tolerates mismatches. The latter is considered a feature allowing the detection of protospacers in phages that have accumulated some mutations. The self versus nonself discrimination is achieved with the help of crRNA sequences complementary to the repeats in the CRISPR arrays. If the gRNA base pairs with a protospacer and a PAM sequence, the complementary DNA is cleaved and the phage infection is prevented. If the gRNA base pairs with a spacer and a repeat (i.e., the CRISPR locus) the complementary DNA is not cleaved, thus the integrity of the CRISPR locus is preserved. Tracing the very short history of the CRISPR/Cas editing offers unprecedented opportunities to appreciate simultaneously how advance science of the 21st century is and how much intellectual effort it takes to decipher a phenomenon starting from single observations. The first publications regarding CRISPR/Cas system reported on array of repeats with unknown function in the late 1980. Over the course of the next decade, advances in sequencing resulted in accumulation of tons of microbial genomes, and eventually such repeats were identified in multiple bacterial and archeal species and their utility for strain identification recognized. In the next decade the Cas genes were discovered, and it was appreciated that the CRISPR spacers are homologous to bacteriophage DNA. Initially and it was proposed that the CRISPR system is probably analogous to the RNA interference (RNAi) mechanisms for antiviral defense in eukaryotes. As knowledge was continuing to accumulate the conceptual principles underlining, the CRISPR/Cas systems were deciphered and its function as adaptive immune system identified. Characterization of the properties of CRISPR/Cas components from various origins resulted in the development the concept of the CRISPR/Cas9 gene editing approach along with relevant tools and applications. Roughly three decades after the initial discovery of the CRISPR repeats, the scientific community is starting to discuss a CRISPR/Cas pipeline of drug discovery, which is expected to transform the capabilities of contemporary medicine (Fig. 3.14).

3.4 RNA INTERFERENCE

RNAi is a complex pathway described in many eukaryotes, associated with defenses against viruses and transposons, which inhibits gene expression by neutralizing specific mRNA molecules (Fig. 3.15). The pathway is activated by the detection of dsRNA molecules in the cell, which are cleaved by an enzyme named Dicer producing short siRNA (small interfering RNA). One of the strands of siRNA is degraded, whereas the other, called a guide RNA (gRNA), complexes with several proteins forming an RNA-induced silencing complex (RISC). When the gRNA strand base-pairs with complementary mRNA, the Argonaute 2/ Ago2 component of RISC degrades the mRNA, and the gRNA is recycled. As a result of the repetitive use of the gRNA, the target mRNA is depleted and thus the corresponding gene silenced. Eukaryotic cells code for microRNAs, which utilize similar mechanism to deplete specific mRNAs or prevent their translation and influence development. RNAi is very efficient defense mechanism against RNA viruses, which generate dsRNA intermediates as a part of their replication cycle. In research settings, RNAi is a powerful tool for altering gene expression in vivo in the context of cell culture or living organisms. Synthetic dsRNAs introduced by transfection or transduction can selectively trigger partial or full suppression of genes of interest, thus influencing cellular and organismal physiology. RNAi is used to downregulate or abolish gene expression for functional studies, to correct metabolic defects or to create research models to dissect complex processes. The technology has been utilized

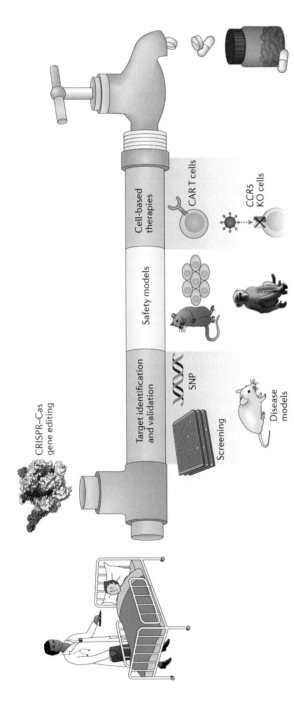

FIGURE 3.14 **Envisioned pipeline of CRISPR/Cas-driven drug design.** Unmet medical needs for numerous diseases and the rapid progress of CRISPR/Cas gene editing can feed into a drug discovery and development pipeline, which leads to improved therapies. The CRISPR/Cas system allows for improved target identification and validation as well as faster generation of safety models. CRISPR–Cas can also be used to develop cell-based therapies, such as chimeric antigen receptor T cells for immunotherapy and C-C motif chemokine receptor 5 (CCR5)-knockout (KO) cells for HIV treatment. CRISPR–Cas-assisted drug discovery will yield innovative therapies and treatment paradigms for patients. SNP, single-nucleotide polymorphism. *Image and figure legend reprinted from Fellmann C, Gowen BG, Lin PC, Doudna JA, Corn JE. Cornerstones of CRISPR–Cas in drug discovery and therapy. Nat Rev Drug Discov 2017;16(2):89–100, with permission from Nature publishing group.*

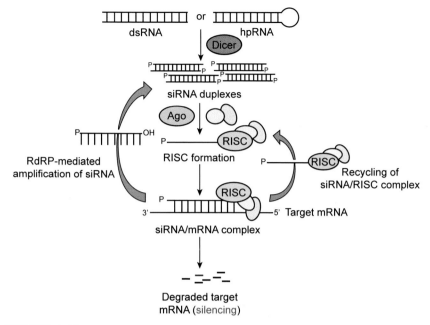

dsRNA or hpRNA

Dicer

siRNA duplexes

Ago

RISC

RISC formation

RdRP-mediated
amplification of siRNA

RISC

Recycling of
siRNA/RISC complex

RISC

3' 5' Target mRNA

siRNA/mRNA complex

Degraded target
mRNA (silencing)

FIGURE 3.15 **Overview of RNAi-mediated gene silencing in eukaryotes.** Double-stranded RNAs or hairpin RNAs (hpRNAs) generate small siRNA duplexes by the action of Dicer. The guide RNA strand binds with Argonaute (Ago) and other proteins to form an RNA-induced silencing complex (RISC). The siRNA/RISC complex then binds the complementary sequence of the target mRNA resulting in the degradation of the target transcript or inhibition of translation. The components of siRNA/mRNA complex can be recycled to the RISC complex or generate siRNA duplexes by the action of RNA-dependent RNA polymerase (RdRP). *Image and figure legend courtesy Majumdar R, Rajasekaran K, Cary JW. RNA interference (RNAi) as a potential tool for control of mycotoxin contamination in crop plants: concepts and considerations.* Front Plant Sci *2017;8:200.* https://doi.org/10.3389/fpls.2017.00200, *CCBY,* https://creativecommons.org/licenses/by/4.0/.

for various applications in biotechnology, medicine, and agriculture. The development of CRISPR/Cas9 gene editing approach offered alternatives to many of the RNAi applications, although each approach has its own strengths and utility. Understanding of the mechanisms of viral anti-RNAi defenses in plants has contributed to designing successful virus-based vectors for molecular farming.

Further Reading

1. Boettcher M, McManus MT. Choosing the right tool for the job: RNAi, TALEN, or CRISPR. *Mol Cell* 2015;58(4):575–85.
2. Bortesi L, Fischer R. The CRISPR/Cas9 system for plant genome editing and beyond. *Biotechnol Adv* 2015;33(1):41–52.

3. Branda CS, Dymecki SM. Talking about a revolution: the impact of site-specific recombinases on genetic analyses in mice. *Dev Cell* 2004;**6**(1):7–28.
4. Gilbertson L. Cre-lox recombination: Cre-ative tools for plant biotechnology. *Trends Biotechnol* 2003;**21**(12):550–5.
5. Kim DH, Rossi JJ. Strategies for silencing human disease using RNA interference. *Nat Rev Genet* 2007;**8**(3):173–84.
6. Murphy KC. Phage recombinases and their applications. *Adv Virus Res* 2012;**83**:367–414.
7. Moore R, Chandrahas A, Bleris L. Transcription activator-like effectors: a toolkit for synthetic biology. *ACS Synth Biol* 2014;**3**(10):708–16.
8. Sternberg SH, Doudna JA. Expanding the biologist's toolkit with CRISPR-Cas9. *Mol Cell* 2015;**58**(4):568–74.
9. Urnov FD, Rebar EJ, Holmes MC, Zhang HS, Gregory PD. Genome editing with engineered zinc finger nucleases. *Nat Rev Genet* 2010;**11**(9):636–46.
10. Warren D, Laxmikanthan G, Landy A. Integrase family of site-specific recombinases A2-Maloy, Stanley. In: Hughes K, editor. *Brenner's encyclopedia of genetics*. 2nd ed. San Diego: Academic Press; 2013. p. 100–5.

4

Viral Tools for Protein Expression and Purification

4.1 OVERVIEW OF PROTEIN EXPRESSION: CHALLENGES AND LIMITATIONS

Proteins are the working horses of living matter fulfilling a broad array of functions and comprising the most diverse group of biopolymers building the cell. Naturally, they are a major focus of research, as well as key molecules in biotechnology and biomedicine. Proteins are expressed in research labs and for commercial applications to (1) study their function, properties, and structure; (2) perform in vitro and in vivo assays to monitor cellular processes; and (3) manufacture antibodies, enzymes, and research tools, etc. As of July 2014, there were several hundred biopharmaceuticals based on recombinant proteins approved for use in the United States and European Union market, with more in the pipelines. In most cases, it is critical to be able to produce large quantities of pure, biologically active protein in a cost-effective manner. In a way, that is almost what viruses are doing during infection while forcing the cell to express the viral genome and preferentially synthesize viral proteins, some of them (capsid proteins) in enormous amounts. Not surprisingly, viral components are frequently used to drive heterologous protein expression in both prokaryotes and eukaryotes.

One can argue that the challenges of recombinant protein production are essentially the same as the challenges faced by viruses during infection. In both scenarios, mRNAs must be fully compatible with the environment of the expressing cell, preferentially synthesized, stable, and efficiently translated. In addition, heterologous protein expression systems must account for sensible purification and detection of the expressed proteins. In that regard, unique viral peptides come handy as protein tags and unique viral proteases as tools for removing them. Conventional expression of recombinant proteins can be accomplished on various scales for research and manufacturing purposes in bacterial cells (*E. coli*), yeast cells (*P. pastoris*), insect cells (baculovirus expression system), mammalian cells, and plant cells or whole plants. Each system has its advantages and disadvantages, which are summarized in Table 4.1. Choices of protein expression systems are driven by multiple factors such as origin of the protein (prokaryotes vs. eukaryotes), protein solubility, posttranslational modifications, protein associations in multicomponent complexes, requirements of chaperones, etc. As the needs of more proteins to be expressed and purified for various purposes increase, approaches for high throughput expression and solubility testing are becoming more and more important, along with the development of cost-effective cell-free expression systems (out of the scope of this chapter).

TABLE 4.1   Overview of Key Model Systems for Protein Expression and Purification

Expression Systems	Advantages	Disadvantages
E. coli	• Rapid (days) • Cheap media • Simple to scale up • Well-characterized genetics	• Limited capacity for posttranslational modifications • Some difficulties to produce soluble and properly folded proteins
Yeast	• Moderately rapid (weeks) • Cheap media • Supports most posttranslational modifications • High folding capacity	• N-linked glycosylation different from mammalian forms • Safety precautions needed for large-scale production due to use of methanol in induction media
Mammalian cells	• Moderately rapid (weeks; transient expression) • Full-scale posttranslational modifications • Proper folding	• Slow (months; stable transfection approach) • Expensive • Hard to scale up
Baculovirus/ insect cells	• Moderately rapid (weeks) • Supports most posttranslational modifications • High folding capacity	• N-linked glycosylation different from mammalian forms • Expensive • Hard to scale up
Plants	• Moderately rapid (weeks; transient expression) • Relatively cheap • Supports most posttranslational modifications • High folding capacity • High potential for scaling up	• Slow (months; transgenic plants) • N-linked glycosylation different from mammalian forms • Different tools have to be developed for different plants

4.2 VIRAL COMPONENTS UTILIZED IN PROTEIN EXPRESSION SYSTEMS

4.2.1 Transcription-Related Viral Components

The success of any system for recombinant protein production is largely based on its ability to produce mRNA molecules that persist in the cell and can be translated efficiently. The rate of transcription is a function of the speed of transcription initiation, which correlates with the strength of the promoter sequences guiding the expression of the gene in consideration. Strong promoters generally present optimal sequence context for their cognate RNA polymerases (RNAPs) and allow highly specific binding and initiation, which make them a great tool for the recombinant expression of soluble nontoxic proteins. On the other hand, proteins with limited solubility or significant toxicity have a bigger chance to be productively expressed from weak promoters. In an ideal world, a strong, inducible, and tightly regulated promoter, allowing "tunable" expression, offers the most experimental flexibility to achieve diverse protein expression goals. Interestingly, despite the fact that viruses infecting all branches of life have strong promoters, not all types of expression systems utilize viral components in a similar fashion. Expression of recombinant proteins in *E. coli* is driven predominantly by strong phage promoter/robust phage RNA polymerase pairs; the yeast expression system does not use viral promoters at all; while mammalian systems use a mix of viral and nonviral promoters driven by cellular RNA Pol II. The baculovirus expression system is completely virus-driven, utilizing viral promoters and host or viral RNA polymerase. Plant expression systems utilize viral promoters and host polymerases in a rather sophisticated setup for gene delivery involving *Agrobacterium tumefaciens*. This subsection will briefly review viral components used in protein expression in *E. coli* and mammalian cells. Individual subsections are dedicated to the baculovirus expression system and plant expression systems due to numerous recent advances.

4.2.1.1 T7 Promoter/T7 RNA Polymerase–Driven Expression

Historically, three different promoter/polymerase pairs have been utilized in recombinant protein expression in *E. coli*: bacteriophage T3, bacteriophage T7, and bacteriophage SP6, among which the T7 system is preferentially used. T7 RNAP is a single polypeptide that recognizes its cognate promoter (5′ TAATACGACTCACTATAG 3′) with very high specificity initiating highly efficient transcription that outpaces the cellular one by half of an order of magnitude. Since the sequence of the T7 promoter is very different from the one for *E. coli* RNAP and does not occur in the *E. coli* genome, the T7 promoter/RNAP/terminator system is very efficient and can transcribe essentially any sequence placed in the right context between the two regulatory signals. Unlike the *E. coli* RNAP, T7 RNAP is not sensitive to rifampicin, thus allowing the genes cloned under the control of the T7 promoter to be exclusively transcribed in the presence of the antibiotic. The utility of the enzyme to mediate robust protein expression was initially demonstrated in the mid-1980s and has been the golden standard for *E. coli* expression of recombinant proteins, known as pET expression system. In addition, T7 RNAP has played a crucial role in advancing our knowledge about RNA structure and function in living matter. In vitro transcription driven by T7 RNAP has enabled the production of countless RNA molecules in a test tube and numerous studies across all domains of life.

The pET expression system utilizes a chromosomal copy of T7 RNAP under the control of the isopropyl β-D-1-thiogalactopyranoside (IPTG)-inducible promoter of the *E. coli lac* operon (Fig. 4.1). IPTG, is a non-hydrolyzable analog of lactose, the natural inducer of the lac operon. The protein of interest is encoded by a pET expression plasmid under the control of a T7 promoter. Since the *E. coli* RNAP does not recognize the T7 promoter, the protein of interest will only be expressed in strains carrying the T7 RNAP gene. This avoids issues related to cloning and maintaining plasmids that encode toxic proteins. Currently, more than 40 different versions of pET plasmids are commercially available, offering various options for cloning, antibiotic resistance, and protein tags, among other features. In a noninduced state, very minimal, if any, recombinant protein is expressed as a result of occasional read-throughs of the T7 RNAP gene by the cellular RNAP. In an induced state, T7 RNAP drives the expression of the gene of interest in a dose-dependent manner based on IPTG concentration. Multiple refinements of both plasmids and *E. coli* strains have been made to adapt the system to meet different protein expression needs (Fig. 4.2). For example, expression of toxic proteins benefits from tight repression of the T7 RNAP gene, which can be accomplished by addition of glucose in the growing medium to take advantage of the catabolic repression property of the *lac* promoter and/or using a cell strain harboring a pLysS plasmid coding for T7 lysozyme, an inhibitor of T7 RNAP. An auto-inducing media has been developed as an alternative to IPTG induction allowing for straightforward and amenable to automatization screening of multiple clones for expression and solubility. The auto-inducing media supports productive expression in high-density cultures, thus improving protein yields several fold. Protein yields can be also improved by using cell strains deficient in RNAse E, which is known to digest mRNAs transcribed by T7 RNAP. The transcription elongation rate of T7 RNAP is much faster than the rate of translation supported by prokaryotic ribosomes, leaving long stretches of mRNA vulnerable to RNAse E. In addition, cell strains carrying plasmids coding for chaperones and/or tRNAs for rare codons have been used as tools to improve folding, solubility, and yields of expressed proteins. Remarkably, T7 polymerase has been found active in other prokaryotic species as well as eukaryotic cells. Eukaryotic systems employing T7 RNAP-coding plasmids or recombinant vaccinia virus expressing T7 RNAP have been developed.

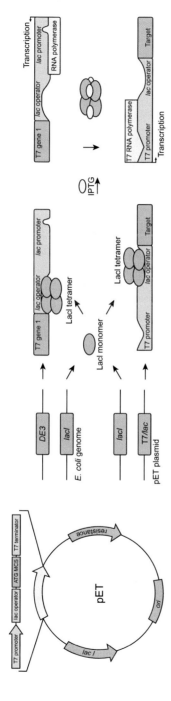

FIGURE 4.1 **Overview of the pET/E. coli expression system.** pET expression vectors (general diagram on the **left**) utilize bacteriophage T7 promoter/terminator pair and lac operator allowing for efficient and tunable expression. In addition of the conventional origin of replication (ori) and resistance marker, many pET vectors carry a copy of the lacI gene coding for the lac repressor and allowing tight transcription regulation. The T7 RNA polymerase coded in the host chromosome is under the control of the lac promoter. The gene of interest is inserted in the multiple cloning site (MCS) positioned in frame with a start codon (ATG). In absence of inducer, lacI is expressed and represses the transcription of both: T7 RNA polymerase and the gene of interest by binding to the lac operator and preventing the respective RNA polymerase from binding there (**middle panel**). In presence of inducer (isopropyl β-D-1-thiogalactopyranoside (IPTG), a nonhydrolyzable analog of lactose), the T7 RNA polymerase is expressed and drives the transcription of the gene of interest (**right panel**). Reprinted from Sørensen HP, Mortensen KK. Advanced genetic strategies for recombinant protein expression in Escherichia coli. J Biotechnol 2005;**115**(2):113–28, with permission from Elsevier.

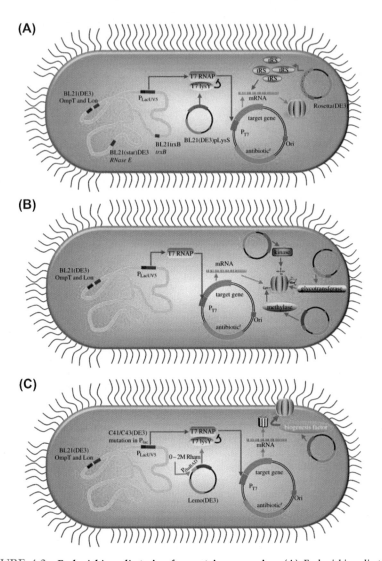

FIGURE 4.2 **Escherichia coli strains for protein expression.** (A) *Escherichia coli* strains widely used in recombinant protein production. In the expression vector, the target gene is under control of the T7 promoter. In the *E. coli* genome, the gene encoding T7 RNA polymerase is under control of the *lac*UV5. The strain BL21(DE3) is deficient in OmpT and Lon proteases. BL21STAR(DE3) is mutated in RNase E, reducing mRNA degradation. BL21trxB promotes the formation of disulfide bonds. In BL21pLysS(DE3), T7 lysozyme is expressed, and the enzyme inactivates any T7 RNA polymerase that may be produced without induction. Rosetta strains are designed to improve the expression of proteins encoded by genes containing rare codons used in *E. coli*. (B) Strategy for expressing a protein with posttranslational modification in *E. coli*. Genes encoding kinases, glycosyltransferases, methylases, ligases, or other modifying enzymes are coexpressed to produce posttranslationally modified proteins. (C) Overview of *E. coli* strains used in membrane protein production. Walker strains (C41(DE3) and C43(DE3)) are commonly used to overcome the toxicity of membrane proteins. In Lemo21(DE3), expression can be tuned by adding different concentrations of L-rhamnose to the culture. Coexpression of membrane protein biogenesis factors may also facilitate the localization of target proteins. lysY, lysozyme; RNAP, RNA polymerase; tRS, tRNA synthetase. *Image and figure legend courtesy Jia B, Jeon CO. High-throughput recombinant protein expression in* Escherichia coli: *current status and future perspectives.* Open Biol *2016;6(8):160196.* https://doi.org/10.1098/rsob.160196, CC-BY, https://creativecommons.org/licenses/by/4.0/.

4.2.1.2 *Mammalian Transient Expression Systems*

Cost-effective expression of large quantities of recombinant proteins in mammalian cells has been a key goal for a long time in the broad field of drug discovery. Only recently, techniques based on transient transfection have advanced enough to make possible a larger scale of protein production that can meet the demands of drug manufacturing. Mammalian expression systems produce proteins with correct folding and posttranslational modifications, which are often problematic when mammalian proteins are expressed in heterologous systems. Among the first protein-based medicines produced in mammalian cells were Epogen/Procrit (Amgen), a preparation of erythropoietin, alleviating anemia; Follistim (Serono), a preparation of follicle-stimulating hormone; Herceptin (Genentech), an anti-Her2 antibody targeting Her2–positive breast cancer, among others. The most utilized promoter driving transient expression in mammalian cells is the human cytomegalovirus (CMV) promoter with an enhancer naturally driving the expression of the immediately early genes of the CMV or human herpesvirus-5. The CMV promoter is a constitutive promoter that is recognized by mammalian RNAP II. Polyadenylation (polyA) signal sequences from various genes are added to mammalian vectors to ensure proper mRNA processing and stability. The most frequently used signals from viral origin are the ones of SV40 early transcription unit and HSV-1 thymidine kinase. The presence of the woodchuck hepatitis virus posttranscriptional regulatory element (WPRE) has been shown to increase protein yields by modulating mRNA stability and export from the nucleus.

Mammalian cells used for expression have been modified to include viral components directing expression plasmids to the nucleus and ensuring their partitioning during cellular division. For example, HEK 293-E cells have been genetically manipulated to express the EBNA-1 (Epstein–Barr nuclear antigen 1) protein, which by virtue of its natural propensity interacts with oriP, the origin of replication of Epstein-Barr virus engineered in the employed expression vector, maintaining the plasmid as an episome in the nucleus as well as directing its proper partitioning to daughter cells during cell division. Viral elements of SV40, bovine papillomavirus, and polyomavirus have been used in a similar fashion. The original HEK 293 cell line is the first cell line transformed by a virus. Human embryonic kidney cells were transfected with sheared DNA genome of adenovirus (Ad5 serotype) using calcium-based transfection and cells continuously dividing in cell culture were selected.

Traditionally, most recombinant proteins purified from mammalian cells are produced commercially using stably transfected cell lines. The process of obtaining well-expressing cell lines is a combination of transient transfection and drug selection steps, and is rather lengthy. Selected cell lines have the gene of interest inserted in random location, thus the level of expression can vary significantly between cell lines even if they were generated as part of the same experimental sequence. In addition, over time some cell lines deteriorate, most likely due to transcriptional silencing of the heterologous gene. As the methods for transient transfection improve, protein purification from transiently transfected cell lines is becoming a viable alternative. Development of cell lines adapted to growth in suspension, cost-effectiveness of transfection agent with high efficiency (polyethylenimine) and availability of methods for production of large amounts of high-quality DNA are among the technological advancements making that possible.

Recombinase-mediated cassette exchange (RMCE) method, initially developed for the purpose of transgenic mouse models, is gaining popularity in the field of mammalian protein expression. The approach employs stably transfected master cell lines harboring a recombination cassette consisting of a reporter gene flanked by recombination target sites such as Lox P (Cre recombination system form bacteriophage P22) or Flp (yeast recombination system). Each master cell line harbors one cassette, inserted in a location tolerating foreign genes and supporting high level of expression. Once good quality master cell lines are produced and selected empirically using the stably transfected cell line approach, the gene of interest is delivered in the cell on a plasmid in the context of the same recombination cassette, i.e., flanked by the same recombination target sites. When a cognate recombinase is expressed in the cell recombination cassette, exchange takes place, effectively replacing the reporter gene of the master cell line with the gene of interest, thus generating new expression cell line. The procedure is commonly referred to as "Cre out" or "Flp out" depending on the system used. The gene of interest cassette plasmid and the gene for the recombination enzyme are inserted in the master cell line via transient transfection or using transduction with viral vectors.

An extension of the master cell line approach pushes protein expression in mammalian cells even further. Stable cell lines with quantifiable protein production have been generated using CRISPR/Cas9-mediated knock-in strategy. The new technology (2017) aims to introduce a gene of interest in the context of polycistronic construct to be expressed from a natural promoter with desirable characteristics. The utility of the method has been demonstrated with the promoters for actin and the ribosomal protein L13A, both driving high, but somewhat different levels of protein expression. The quantifiable feature is accomplished by including a fluorescent protein (RFP or GFP) in the construct, whereas selection is driven by a zeocin marker. The polycistronic construct contains all three genes: fluorescent marker, gene of interest, and zeocin-resistance marker connected by sequences of self-cleavable peptides p2A (Section 4.2.2) and flanked by locus-specific sequences allowing for CRISPR/Cas 9-driven integration in the target locus. The successful construct integration results in simultaneous expression of four genes driven from the same natural promoter. Four individual proteins are released as a consequence of the action of the self-cleaving peptides: the natural protein whose promoter is used for the expression; fluorescent protein reporter whose expression can be quantified spectrophotometrically, gene of interest and the zeocine-resistance marker. Since the zeocin-resistance marker is coded last in the polycistronic construct, any selected clone resistant to zeocin will express all four proteins.

Several systems based on mammalian viruses (retroviruses, vaccinia virus, Semliki Forest virus) are being used to deliver heterologous genes into mammalian cells via transduction; however, the approach is rarely used for the sole purposes of protein expression outside a research lab due to cytopathic effects and safety concerns. Baculoviruses are gaining strength as a vehicle to deliver heterologous genetic material into mammalian cells. Although they easily infect mammalian cells, no productive infection takes place due to promoter incompatibility. At this moment, baculoviruses are considered more of a tool for gene therapy rather than protein expression, and most recent advances in that respect are described in Chapter 9. Very likely they will become a force in mammalian protein expression too.

4.2.2 Translation-Related Viral Components

Efficiency of protein expression has been shown to be influenced by sequences in the 5' and 3' UTRs (untranslated regions) of viral mRNA. In addition to the conventional elements required for efficient interaction between mRNA and the ribosome such as Shine-Dalgarno sequence in prokaryotes, Kozak sequence, internal ribosome entry sites (IRESs) and polyA signal in eukaryotes, they often contain translational enhancers. Translational enhancers have been identified in both 5' and 3' UTRs of viral mRNAs in both prokaryotic and eukaryotic viruses. Prokaryotic expression vectors most frequently contain the 5' leader of the bacteriophage T7 gene 10, also known as Epsilon (*enhancer of protein synthesis initiation*) element. Plant expression vectors frequently contain the translational enhancers from potato virus X or tobacco mosaic virus, both enhancing CAP/polyA-dependent translation initiation. The mechanisms of action of translation enhancers are not well understood and their abundance and diversity is just beginning to be appreciated.

IRES are among the best-understood group of translation-related UTR sequences that is slowly but surely gaining ground as a component of vectors for expression of recombinant proteins, as well as vectors for gene therapy. The first IRES sequences were described in poliovirus and encephalomyocarditis virus (EMCV). IRES elements vary widely in length, secondary structure, and mechanism of action. Some can initiate translation without assistance of any conventional translation initiation factors, others require a minimal subset of them (Fig. 4.3). IRES sequences are used most frequently in polycistronic expression vectors coding for multiple proteins. Usually the translation of the first protein is CAP-driven, whereas the translation of the second and each subsequent one is driven by unique IRES sequences, allowing efficient concurrent expression (Fig. 4.4). The IRESite database keeps track of identified IRES sequences and experimental data demonstrating their activity.

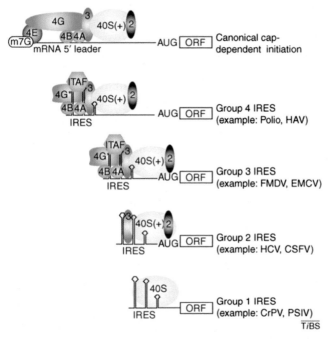

FIGURE 4.3 **Internal ribosomal entry site (IRES) groups and their requirement for translation factors.** Conventional translation initiation in eukaryotes employs 5′ CAA, 3′ polyA tail and plethora of translation initiation factors (**top panel**) working to position the ribosome and the initiator tRNA^{Met} at the initiation codon. IRES elements allow translation to be initiated on CAPless mRNAs and with a variable number of eukaryotic translation initiation factors (eIFs). Some viruses replace conventional eIFs with ITAFs (IRES-transactivating factors). IRES group I binds to the ribosome directly without assistance of any eIFs or ITAFs and does not require an initiator tRNA (*CrPV*, cricket paralysis virus; *PSIV*, Plautia stali intestine virus; *TSV*, Taura syndrome virus). IRES group II binds to the ribosome employing a subset of conventional eIFs and tRNA^{Met}, and does not require any ITAFs (*CSFV*, classical swine fever virus; *HCV*, hepatitis C virus). Group III requires a subset of conventional eIFs, tRNA^{Met}, and ITAFs (*EMCV*, Encephalomyocarditis virus; *FMDV*, foot-and-mouth disease virus). Group IV requires the same factors as group III and additional cell extracts from specific cell lines to function in vitro in the context of rabbit reticulocyte lysate (*HAV*, hepatitis A virus). *Reprinted from Kieft JS. Viral IRES RNA structures and ribosome interactions. Trends Biochem Sci 2008;33(6), 274–83, with permission from Elsevier.*

A viable alternative of IRES (or complementary tool) in polycistronic vectors are the P2A self-cleaving peptides (Fig. 4.4B). The first one was isolated from foot-and-mouth disease virus not so long ago. The name P2A comes from the location of the peptide at the end of protein P2A. Currently, the ViralZone/ExPASy Bioinformatics Resource Portal lists a total of 18 P2A self-cleaving peptides from 5 different families of (+) ssRNA and dsRNA viruses. The peptides are very short (18–35 aa) and share a well-conserved C-terminal sequence ExNPGP, where "x" can be any amino acid. The cleavage event takes place between the C-terminal glycine and proline, leaving most of the peptide sequence attached to the upstream protein and an extra proline attached to the downstream protein. In many cases, the peptide leftovers are trimmed by cellular proteases. The P2A peptides work in vivo and in vitro in eukaryotic systems only; however, their activity varies between cell types. Their mechanism of action is rather intriguing. Contrary to what their name suggests, no

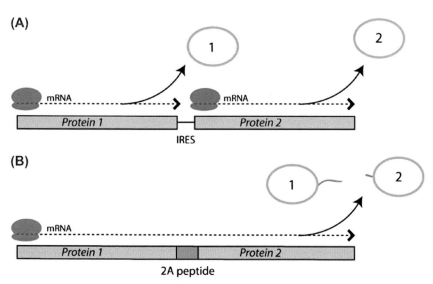

FIGURE 4.4 **Application of internal ribosomal entry site (IRES) and p2A self-cleaving peptides in polycistronic constructs.** Polycistronic constructs for the simultaneous expression of two heterologous proteins (Protein 1 and Protein 2). (A) CAP/IRES-mediated translation. The polycistronic mRNA undergoes two separate translation initiation events. Protein 1 translation is initiated via Cap-dependent mechanism, whereas protein 2 translation is initiated via IRES-dependent mechanism. (B) CAP/2A peptide-mediated translation. The polycistronic mRNA undergoes one translation initiation event using CAP-dependent mechanism and one "ribosome skipping" event resulting in stop codon-less termination for protein 1 and perceived initiation at the C-terminal end of the P2A peptide. *Image courtesy Sainsbury F, Benchabane M, Goulet MC, Michaud D. Multimodal protein constructs for herbivore insect control.* Toxins *(Basel) 2012;4(6):455–75.* https://doi.org/10.3390/toxins4060455, CC-BY, https://creativecommons.org/licenses/by/4.0/.

actual cleavage event takes place, instead, perceived cleavage is achieved by unconventional stop codon-less ribosome termination event, which has been described as ribosome "skipping," ribosome "stop-and-go" or "stop-and-carry" mechanism. When the ribosome is translating the C-terminal sequence of the P2A peptide, the peptidyl transferase activity is inhibited presumably by the ExNPGP motif and cannot form a peptide bond between the glycine at the very C-terminus of the polypeptide being synthesized and the upcoming proline, effectively resulting in a termination event and the release of the already synthesized polypeptide chain. Since the A site of the ribosome is occupied from tRNA carrying proline and not a termination factor, the ribosome does not dissociate. Instead, it continues to work in an elongation mode (adding new amino acids to the growing polypeptide with a proline at its N-terminus) until it reaches stop codon. Collectively, when comparing the coding mRNA template and the resulting polypeptide products, one can easily assume that the resulting proteins are a product of cleavage event, which explains the name of the P2A self-cleaving peptides. Regardless of the discrepancy between their name and their mechanism, the P2A peptides are a very useful trick/gadget for both viruses and expression vectors allowing efficient utilization of coding potential.

Similar to viruses, expression vectors have a limit of coding capacity, which is easily reached when multiple proteins need to be expressed simultaneously; for example, when virus-like particles are to be produced or a multisubunit enzyme is to be purified. Since most of the proteins in nature work as part of multicomponent complexes, the need of creative strategies for maximizing the coding capacity of expression vectors will only increase. Viruses often utilize proteolytic cleavage to generate individual proteins from coded polyproteins. In some cases, employed proteases are a part of the polyprotein itself, in others, they are independently coded viral or host enzymes (Fig. 4.5). The depicted strategies are currently used to secure processing of polyproteins or to ensure equimolar expression of multiple polypeptide chains.

4.2.3 Molecular Tools Facilitating Protein Detection and Purification

Protein tags are a useful and convenient tool for improving solubility of recombinant proteins, streamlining protein purification, and allowing an easy way to track proteins during protein expression and purification. In addition, protein tags are a useful tool for tracking proteins and processes directly in live cells using microscopy or indirectly using Western blot, immunoprecipitation, or immunostaining. Viral proteins are good candidates as a source of protein tags since they are likely to be unique, thus minimizing possible issues with cross-reactivity of antibodies used

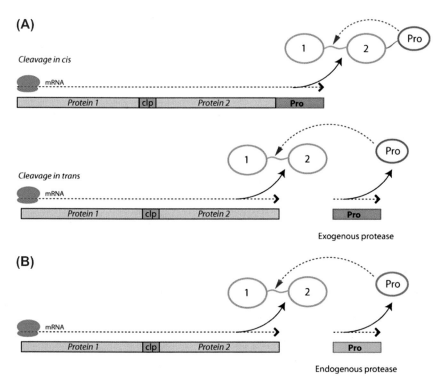

FIGURE 4.5   **Viral mechanisms of polyprotein proteolysis.** A polyprotein consisting of two hypothetical proteins, Protein 1 and Protein 2, are expressed from a single construct. The polyprotein includes a cleavable linker peptide (clp) (in green) between Protein 1 and Protein 2, which is posttranslationally cleaved producing two mature proteins. (A) Exogenous protease-mediated cleavage. The polyprotein is cleaved off by a recombinant protease expressed as part of the polyprotein or by an enzyme coded on a separate construct (B) Endogenous host cell protease-mediated cleavage. *Black arrows* on panels (A and B) indicate the direction of translation. *Red arrows* point to protease cleavage sites. *Image courtesy Sainsbury F, Benchabane M, Goulet MC, Michaud D. Multimodal protein constructs for herbivore insect control.* Toxins (Basel) *2012;4(6):455–75.* https://doi.org/10.3390/toxins4060455, *CC-BY,* https://creativecommons.org/licenses/by/4.0/.

to visualize them as well as copurification of cellular proteins. Tags can be attached to recombinant proteins N-terminally or C-terminally, depending on experimental needs and the properties of the protein of interest. On many occasions, tag addition may result in partial or complete loss of function warranting the need for complementation studies and side-by-side analysis of tagged and untagged protein versions. Among the most widely used protein tags, several are from viral origin: (1) T7 tag (MASMTGGQQMG), composed of the N-terminal 11 residues of the major T7 capsid protein, gp 10; (2) HA tag (YPYDVPDYA), encompassing residues 98–106 of the major surface protein of influenza virus, hemagglutinin and (3) HSV tag (SQPELAPEDPED), encompassing residues 290–300 of glycoprotein D, a major surface protein of herpes simplex virus type 1 (HSV-1). In special cases, tandem tags can be engineered allowing different steps in purification/detection to be carried out with the utility of individual tags. Most frequently, tandem tags are employed for purification of proteins with low abundance and/or to ensure use of optimal antibodies for different experimental methods.

When needed, protein tags are removed by specific proteases, whose target sites are engineered as part of the expressed protein. Proteases of viral origin are frequently used due to their high activity and specificity. The 3C protease of human rhinovirus 14, recognizing the LEVLFQ/GP sequence (and generating a cut at the slash mark), is among the most popular ones. It is commercially available as a GST-fusion recombinant protein produced in E. *coli* under the name PreScission protease (Amersham), which is straightforward to remove from protein preparations using affinity chromatography. TEV protease from tobacco etch virus, recognizing the ENLYFQ/G sequence, is also a popular choice in part due to availability of protocols for straightforward in-house expression and purification. A protease with a recognition site very similar to the TEV protease is the TVMV protease (ETVRFQG/S) from tobacco vein mottling virus.

4.3 EXAMPLES OF VIRUS-DRIVEN PROTEIN EXPRESSION SYSTEMS

4.3.1 Baculovirus Protein Expression in Insect Cells

Baculoviruses are widely used to express eukaryotic proteins that are difficult or impossible to produce otherwise. The expression system has been continuously improved to address various challenges and to make the process user-friendly requiring minimal specialized expertise related to baculoviruses themselves. Recent developments in the field are seeking to extend the applications of the baculovirus expression system beyond conventional protein production with the aim to tackle the complexity of living matter by enabling expression of multicomponent protein complexes and to improve the efficiency of producing vectors for gene therapy (Chapter 9). The discussion of the baculovirus expression system will begin with a brief overview of baculovirus molecular characteristics, continue with a review of the evolution of the protocols for generating recombinant baculoviruses, and finish with a summary of current issues and outlook for the future. In addition to their use as a tool for protein expression, baculoviruses are widely used as an agent for insect biocontrol. Aspects of baculovirus biology related to biocontrol are discussed in Chapter 7.

Baculoviruses are rod-shaped, enveloped dsDNA viruses with genomes ranging between 80 and 180 kilobases. More than 50 different baculoviruses have been described infecting more than 600 insect species. A key feature of baculoviruses is the formation of two types of virions: budded virions (BVs) and occluded virions (ODVs/occlusion-derived virions) (Fig. 4.6). BVs are released as individual particles through budding and spread the virus from cell to cell within the infected animal, whereas the ODVs are embedded in occlusion bodies in the nuclei of the infected cells and are released to the environment when the cells are lysed and the host dies. Depending on the species, occlusion bodies can contain one

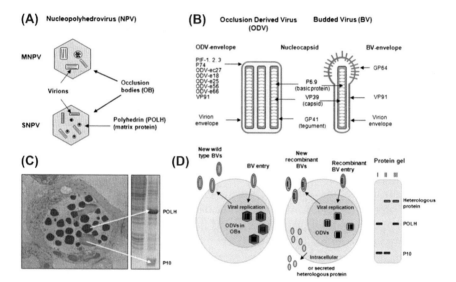

FIGURE 4.6 **Baculovirus features relevant to protein expression.** Nucleopolyhedroviruses survive outside their host in the form of viral occlusion bodies (OBs) that consist of a polyhedrin (polh) matrix in which the enveloped occlusion-derived virions (ODVs) are embedded. The ODVs contain single (SNPV) or multiple (MNPV) nucleocapsids (A). Apart from ODVs that spread the infection between insects and infect midgut cells, a second virion type, the budded virus (BV) spreads the infection from cell to cell in the insect body (B). Phenotypically, BVs and ODVs are different, but their genome is the same. During a virus infection, two proteins, POLH that forms the matrix of viral OBs and P10 that forms fibrillar structures (C, left panel), are made in very high amounts, as seen in a protein gel (C, right panel). Neither of these highly abundant proteins is required for infection of cell cultures, and therefore their promoters are ideal to drive the expression of foreign genes in the baculovirus expression system. (D) The left panel shows the infection of insect cells with wild-type (WT) BV, the middle panel with a recombinant virus where the *polh* promoter drives the expression of a foreign gene, resulting in the absence of OBs, and an intracellular or secreted heterologous protein. The right panel shows a schematic of a protein gel for WT-virus-infected (I), and polh (II) and p10 (III) promoter-based recombinant baculovirus–infected insect cells. *Image and figure legend reprinted from van Oers MM. Opportunities and challenges for the baculovirus expression system. J Invertebr Pathol 2011;107(Suppl.):S3–15, with permission from Elsevier.*

or multiple virions occluded in thick protein coat. *Autographa californica* multicapsid nucleopolyhedrovirus (AcMNPV) is most frequently used for protein expression in tissue culture, whereas *Bombyx mori* nucleopolyhedrovirus (BmNPV) is currently being developed as a system for protein expression in caterpillars of the silk worm (*Bombyx mori*). The discussion bellow is limited to AcMNPV; however, most features of AcMNPV are applicable to baculoviruses as a group since they share a large core of conserved genes relevant to virus propagation. AcMNPV uses both strands of its dsDNA genome to code for proteins in nonoverlapping ORFs (open reading frames). Gene expression follows the conventional phase pattern of dsDNA viruses with the addition of a very late phase signified by the expression of polyhedrin (polh) and p10 genes coding for proteins associated with the formation and the release of the occlusion bodies. Both genes are highly expressed by virally encoded polymerase. Polh and P10 proteins are not essential for the production of budding virions and thus their coding sequence can be replaced with heterologous DNA sequences coding for proteins of interest. Interestingly, AcMNPV can enter mammalian cells; however, promoter incompatibility prevents productive replication. When mammalian promoters are introduced in the system, AcMNPV can be utilized as a vector for gene delivery.

The success of the baculovirus expression system stems from its ability to produce large amounts of recombinant protein driven from the polh promoter, the compatibility of mammalian and insect posttranslational modification systems, and the clinical compatibility between baculovirus protein preparations and the human body. Since AcMNPV does not productively propagate in human cells and no human pathogens are known to propagate in the insect cells utilized by the system, the danger of contaminating human pathogens is minimal if not nonexistent.

For the purpose of recombinant protein production, AcMNPV is propagated in insect cell lines isolated from Spodoptera frugiperda (fall armyworm) and Trichoplusia ni (cabbage looper), known as Sf9, Sf21, and High Five, respectively. All three cellular types can be cultured, immobilized on hard surface or in suspension. The infection is accomplished by recombinant baculoviruses coding for the protein of interest. The typical recombinant baculovirus contains heterologous cDNA sequence under the control of the polh promotor in the natural location of the *polh* gene. The strength of the *polh* promoter ensures high level of mRNA production and in most cases high level of protein expression.

Initially, recombinant baculoviruses containing heterologous DNA sequences under the control of the *polh* promoter were generated taking advantage of natural mechanisms of homologous recombination including the flanking sequences of the *polh* gene (Fig. 4.7). The gene of interest was cloned on a transfer plasmid in the context of the upstream and downstream flanking sequences of the *polh* gene and the construct propagated

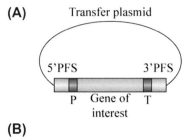

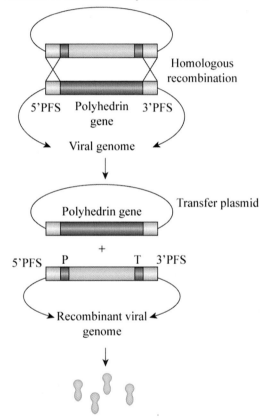

FIGURE 4.7   **First generation baculovirus expression vectors.** The first recombinant baculoviruses were generated by taking advantage of the natural propensity of the virus for homologous recombination. Transfer plasmids (A), coding for the protein of interest in the polyhedrin locus, flanked by the natural polyhedrin sequences, were transfected into insect cells, which were then superinfected with wild-type (WT) baculovirus. Homologous recombination took place resulting in recombinant viral genomes (B), which were packaged as recombinant viruses. The viral progeny was a mix of WT and recombinant viruses (only recombinant viruses are shown for simplicity). The latter were selected based on plaque morphology. *P*, promoter; *PFS*, polyhedrin flanking sequences; *T*, terminator.

and purified from *E. coli*. Next, a cotransfection of the transfer plasmid and wild-type AcMNPV infectious DNA was performed to allow homologous recombination to take place. Recombinant baculoviruses arise as a result of a double crossover event that exchanges the genetic material in the *polh* locus between the transfer plasmid and the viral genome. The lack of polh expression subtly alters the morphology of the viral plaques, allowing for discrimination between wild type and recombinant progeny. Naturally, the frequency of recombinant viruses is very low (~0.1%) and the subtle differences in plaque morphology required experience and expertise to ensure proper identification of the desired plaques. The process of identifying recombinant viruses was gradually streamlined by modifying both the transfer plasmid and the cotransfected baculovirus genome resulting in incorporation of selection markers and eliminating the ability of the non-recombinant viruses to propagate. Extensive details and recent advances in protocols for generating recombinant baculoviruses are reviewed periodically in the specialized literature and can be found elsewhere. In brief, the selection of recombinant viruses was improved by introducing restriction site(s) in the genome of AcMNPV and using linearized infectious DNA for recombination. This resulted in decreased background of wild-type plaques and improved the frequency of recombinant baculoviruses to 20%. Furthermore, the above strategy was complemented with partial deletion of orf1629, a gene essential for AcMNPV propagation, increasing the frequency of recombinants to 95%. The screening process itself was improved by introducing a LacZ marker in the baculovirus genome that allowed blue/white discrimination between recombinant and non-recombinant plaques. Plaque assays were subsequently eliminated from the process of generating recombinant baculoviruses by utilizing bacterial artificial chromosomes (BACs) containing baculovirus genomic DNA (Fig. 4.8) and transposon-driven integration of the heterologous gene. Recombinant BAC clones were identified in *E. coli* using white/blue (Lac Z-based) colony selection and the purified DNA was transfected into insect cells to produce 100% recombinant virus preparation. In subsequent years, many variations of the above approaches were reported that further refined the baculovirus expression system and sought to take advantage of new tools in the fast-changing landscape of molecular biology and provide solutions to new goals in protein expression. Plaque assays for quantitation of produced baculovirus stocks were replaced by immunological and PCR-based methods that are amenable to automation.

The baculovirus system offers capabilities for simultaneous expression of multiple proteins working together as multicomponent complexes (Fig. 4.9). Initially, this was demonstrated by infecting insect cells with multiple recombinant viruses each coding for an individual component, which is a highly inefficient approach since there was no meaningful way to select cells expressing all the components simultaneously.

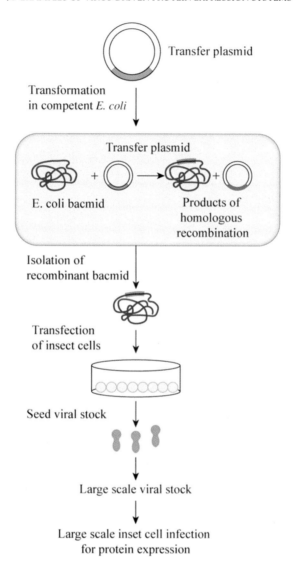

FIGURE 4.8 **Bacmid-based approach to generation of recombinant baculoviruses.**
Bacmids are based on the concept of BAC (bacterial artificial chromosome) vectors, which
are plasmids based on the origin of replication of the *E. coli* F′ (fertility) plasmid and thus
have strict partitioning during cellular division. BACs have a huge coding capacity and anti-
biotic-resistance marker and are propagated in *E. coli* in a similar manner to any plasmids.
Bacmids are essentially BAC clones that code for a baculovirus genome and function as a
shuttle between *E. coli* and insect cells. Transfer plasmid coding for the gene of interest is
transformed in *E. coli* harboring a bacmid. Recombination event takes place via homologous
recombination, transposition, or site-specific recombination, resulting in a recombinant bacmid. Bacmid DNA is isolated from *E. coli* and transfected in insect cells giving rise of recombinant baculoviruses expressing the gene of interest.

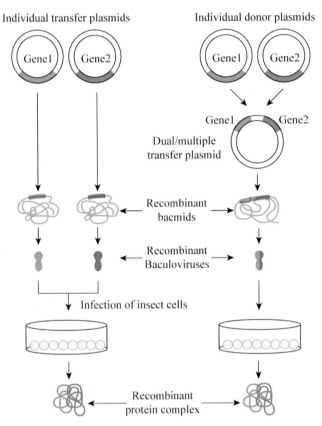

FIGURE 4.9 **Strategies for expression of multicomponent protein complexes using baculoviruses.** Multiple proteins can be coexpressed in insect cells via coinfection with multiple baculoviruses, each expressing an individual protein (left) or via infection with recombinant virus coding for two or more genes of interest (right). In the latter case, donor plasmids coding for each individual gene of interest are assembled in multigene transfer plasmid before bacmid recombination takes place. Both strategies can be combined if multiple genes are to be coexpressed and the coding capacity of the transfer plasmid is reached.

Later, dual and modular expression vectors were introduced coding for two or more heterologous genes, each expressed as an individual ORF. A newer approach in coexpression is assembling genes in a polycistronic mRNA allowing Cap-dependent translation initiation for the first gene in the array and IRES-dependent initiation for the subsequent ones or expression of several proteins as one polyprotein, a strategy widely employed in the viral world. Individual polypeptide chains are generated by proteolytic cleavage at engineered protease sites or by self-cleaving peptides derived from picornaviruses. Technologies for efficiently combining multiple genes on one bacilovirus vector are of high interest for the purpose of expression of multicomponent complexes in which individual subunits depend on each other to achieve proper folding, stability, and full-scale function. Among the latest advancements is the MultiBac system, which offers modular cloning and assembly of a single expression vector coding for several proteins. Interestingly, the MultiBac system employs elements from several different viruses: origin of replication from bacteriophage R6Kγ; polh and pol10 promoters from baculovirus, Cre/LoxP site-specific recombination system from bacteriophage P2, HSV-1 tk poly A signal and 3′ UTR (untranslated region), SV40 poly A signal and 3′ UTR.

As with any other expression system, the actual outcome from each expression event is highly dependent on the protein being expressed. In general, under optimal conditions, smaller cytoplasmic proteins have the greatest levels of expression (at least 100 mg/L cell culture), followed by secretory proteins measuring on the scale of tens of mg/L of cell culture, whereas the expression of membrane proteins is challenging. Baculovirus cell surface display approach is considered a step forward in the expression of membrane proteins. The approach calls for inserting the gene of interest in the context of gp64, an essential viral gene with function in viral infectivity. A second copy of the gp64 is placed under the control of the *polh* promoter and the heterologous DNA inserted right after the signal directing the protein for membrane insertion. Both mammalian and insect signal sequences were found to efficiently guide proteins through the secretory pathways of the insect cells. The environment of the insect cells supports the formation of disulfide bonds in a conventional way, as well as most of the posttranslational modifications. One exception is N-linked glycosylation, which works differently in mammalian and insect cells. Attempts to resolve that issue were made by generating recombinant baculoviruses coding for mammalian N-glycosylation enzymes in loci different from the *polh* locus or by engineering the host insect cells to express the relevant enzymes. SweetBac, a modified version of the MultiBac system, aims to improve *N*-glycosylation of mammalian proteins by expressing *N*-acetylglucosaminyltransferase II from *C. elegans* and the bovine β1,4-galactosyltransferase I simultaneously with the protein of interest.

Multiple gene manipulations have been attempted to improve expression yields by removing protease genes, inserting chaperones, or inserting transcription enhancers. Another recent development in the field is the engineering of virus-free expression approach using mutants lacking a gene essential for replication. Replication-deficient viruses with cloned gene of interest are propagated in permissive cell line to produce recombinant viral stocks. The protein expression is carried into nonpermissive cell line effectively eliminating the possibility for virus contamination of the purified protein, as well as removing purification step. It is hoped that this approach will improve the social acceptance of protein-based medicines produced using virus-driven manufacturing. An alternative to AcMNPV-driven expression is the silk worm–based system employing

BmNPV (Bombix mori nuclear polyhedrosis virus). Its advantages include lower cost, utilization of serum-free media, higher viral titers, higher levels of protein expression and higher solubility of recombinant proteins, less safety concerns due to the fact that BmNPV is highly specific unlike AcMNPV, availability of tons of knowledge and infrastructure directly related to manufacturing. The baculovirus expression system is to continue to evolve and advance as its applications for production of VLPs (viruslike particles), vaccines, and vectors for gene therapy in mammalian systems set new goals for the field.

4.3.2 Plant Molecular Farming: Principles and Outlooks

Plant molecular farming is a relatively young field whose tremendous advances were brought in the public eye recently in conjunction with the Ebola outbreak (2014–16) and the most promising tool to fight the disease, ZMapp. ZMapp is a biopharmaceutical preparation composed of three monoclonal antibodies produced in tobacco by molecular farming. The drug was used on the ground in the heat of the outbreak in West Africa before classical clinical trials were completed and its full evaluation is still ongoing at the time of writing this chapter.

Plant molecular farming can be generally defined as a large-scale production of recombinant proteins for the purpose of their use as biotechnological products with various applications. While different experts in the field consider a broad set of scientific breakthroughs as first examples of molecular farming, they agree that the approach holds a tremendous potential for cost-effective, fast, and safe production of large amounts of recombinant proteins to be used as biopharmaceuticals, veterinary products, research reagents, food supplements, etc. Molecular farming as a field deals not only with expression of recombinant proteins but also with all aspects relevant to growing and harvesting of the actual plants producing them, as well as approaches to protein extraction, storage, and delivery forms to the final consumer. In parallel with scientific advancements, progress has been made in relevant policies defining good manufacturing practices and testing protocols for new products. Currently, more nonpharmaceutical recombinant proteins, product of molecular farming, are reaching our lives than actual medicines due to the vast differences in safety testing protocols. An extensive list of nonpharmaceutical preparations can be found in article 10 from the recommended further readings, whereas a list of molecular farming products in clinical testing can be found in article 7 from the recommended further readings.

Currently, three principal approaches are being utilized to express recombinant proteins in plants: (1) stable transformation; (2) plant cell suspension cultures; and (3) viral vector–mediated transient expression in whole plants.

The first two approaches are largely based on the transformation capabilities of *Agrobacterium tumefaciens* and the associated Ti (tumor-inducing) plasmid, a system with proven utility for direct transfer of foreign genetic material into plant cells and its integration into the plant genome resulting in subsequent expression of the transgene for the purpose of recombinant protein production or cultivating traits of interest. Since the system capitalizes on the natural mechanisms of plant host/bacterium interactions, the involvement of viral components is limited to usage of viral promoters, most frequently the 35S promoter of cauliflower mosaic virus (CaMV). CaMV is a dsDNA virus that replicates with an RNA intermediate and infects predominantly species from the Brassicaceae (cauliflower, cabbage, radishes, among others). The 35S CaMV promoter is constitutive and works very efficiently in most dicots, relatively poorly in monocots (especially cereals), and surprisingly in some nonplant species ranging from *E. coli* to fungi and vertebrate animals. The latter findings are somewhat controversial and the 35S CaMV remains widely used, although the current trends point toward the need of strong inducible promoters to balance challenges of plant growth and protein expression. Ideally, it would be beneficial if transgenes are induced just before harvest limiting possible interference with plant growth and development.

Viral vectors for transient expression of recombinant proteins capitalize on the natural host range of the parent viruses and their ability to spread systemically. Viruses utilized in transient protein expression include tobamoviruses, for example, TMV; Potexviruses, for example, Potato Virus X (PVX); Tobraviruses, for example, tobacco rattle virus (TRV); Geminiviruses, for example, bean yellow dwarf virus (BeYDV); Comoviruses, for example, cowpea mosaic virus (CPMV).

Over the years, vector capabilities have been growing as the understanding of the basic biology of each virus has increased. Viral vectors can be classified in two major groups: full virus vectors and deconstructed viral vectors, also called first and second generation viral vectors. Since different plant species have been explored as protein-expressing machines, multiple viral vector systems have been developed in parallel. Major recent breakthroughs in the field are increasing the size of expressed proteins, developing approaches for delivery of virus genetic information using *Agrobacterium tumefaciens* as a vehicle, and tandem expression of multiple proteins.

First generation viral vectors essentially deliver the gene of interest as part of viral infectious nucleic acid or viral particle under the control of a strong promoter. The preparations are sprayed over the host plant in a mixture with abrasive particles. The resulting microabrasions present open doors allowing infection to take place, producing many infectious particles, which spread throughout the plant (Fig. 4.10). The expression of the recombinant protein was found to be low due to the long time needed

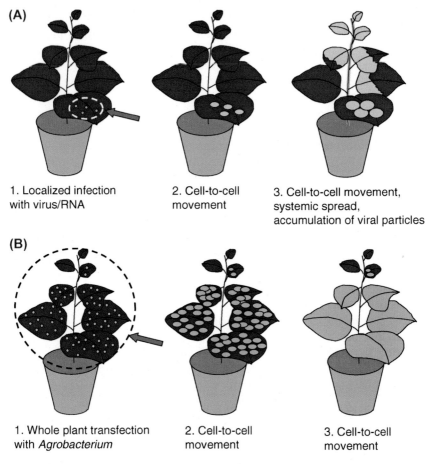

FIGURE 4.10 **Modes of introduction of viral vectors for protein expression in whole plants.** First generation vectors for recombinant protein expression in plants are based on whole viruses able to systemically spread through the entire plant when introduced by localized infection (A). The resulting expression is asynchronous in different plant sections and the coding potential of the vector is limited to 1 kB. Second generation vectors (deconstructed vectors) have larger coding capacity and are delivered to plants via whole plant transfection utilizing the capabilities of Agrobacterium tumefaciens (B). Protein expression is synchronous resulting in high yields. *Reprinted from Gleba Y, Klimyuk V, Marillonnet S. Viral vectors for the expression of proteins in plants.* Curr Opin Biotechnol 2007;*18(2):134–41, with permission from Elsevier.*

for the entire plant to become infected, instability of the viruses carrying large-size heterologous genetic material, and the diversion of protein synthesis resources toward production of capsid proteins. These difficulties led to the development of a second generation of viral vectors, so-called deconstructed viral vectors, which contain viral components relevant to

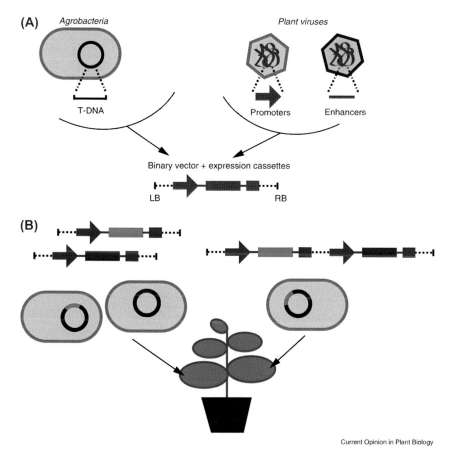

FIGURE 4.11 **Overview of plant transient protein expression system employing second generation vectors.** Current transient protein expression systems in plants combine the advantages of *Agrobacterium tumefaciens* and plant viral regulatory elements to assemble a binary vector system for high-yield protein expression (A). Simultaneous expression of multiple proteins of interest is achieved by coinfiltration of multiple Agrobacterium cultures harboring individual binary vectors or single Agrobacterium culture harboring single vector coding for multiple expression cassettes (B). *Image courtesy Sainsbury F, Lomonossoff GP. Transient expressions of synthetic biology in plants.* Curr Opin Plant Biol *2014;19:1–7.* https://doi.org/10.1016/j.pbi.2014.02.003, *CC-BY,* https://creativecommons.org/licenses/by/3.0/.

the actual protein expression, whereas other functionalities necessary for productive expression are provided in trans (Fig. 4.11). The utility of the deconstructed vector system builds on (1) high expression levels driven by plant viral elements; (2) high transfection efficiency driven by Agrobacterium; (3) the capabilities of plant cells to carry eukaryotic-type posttranslational modifications; and (4) low production cost. The systemic spread of recombinant viruses delivering the gene of interest to many cells is currently being accomplished by combining viral vectors and the classic *Agrobacterium* system for transformation into one approach, known as agroinfiltration or magnifection (when executed on a large scale in industrial settings). During agroinfiltration, viral vectors are introduced in *Agrobacterium tumefaciens* as plasmids and bacterial suspension is grown. Depending on the purpose of the experiment, plant leaf disks, whole leaves, or entire plants are exposed to the bacterial suspension via needle-less syringe injection (on the scale of disks or individual leaves) or by applying a vacuum to a dipping chamber (Fig. 4.12) with whole plants. In both cases, the bacterial culture enters the openings of the stomata and eventually "floods" plant tissues effectively allowing all or almost all cells to be transformed, thus delivering the viral vector inside each individual cell, which then expresses the gene of interest at high yield. Since plants are equipped with defense-silencing systems, transgenic plants expressing

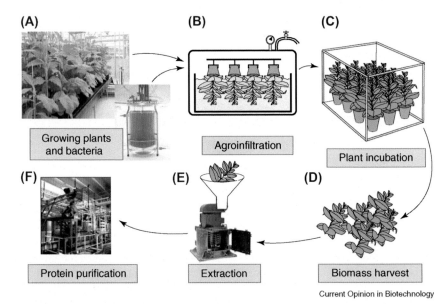

Current Opinion in Biotechnology

FIGURE 4.12   **General scheme for recombinant protein production in plants using agroinfiltration/magnifection.** The recombinant protein expression cycle takes 6–10 days starting with agroinfiltration/magnifection of grown plants and Agrobacterium culture (A). Magnifection is accomplished by submerging plants in bacterial culture harboring plasmid vectors coding for gene of interest and subjecting them to vacuum pulse to force the bacterial culture in (B). Plants are incubated for several days (C) and harvested (D) by strictly controlled protocol designed to prevent the release of genetically engineered bacteria into the environment. Protein extraction (E) and purification (F) are performed using the equipment and infrastructure already in place for protein purification processes related to stably transformed plants. *Reprinted from Gleba Y, Klimyuk V, Marillonnet S. Viral vectors for the expression of proteins in plants.* Curr Opin Biotechnol 2007;*18*(2):134–41, *with permission from Elsevier.*

silencing suppressor proteins are used for protein expression or alternatively a mixed bacterial suspension containing cells coding for the silencing suppressor and cells coding for the gene of interest is used for agroinfiltration. Not surprisingly, the silencing suppressor proteins used in molecular farming are from viral origin in a way showcasing elaborate products of plant–virus coevolution. Among the most frequently employed silencer suppressors is the P19 suppressor from tomato bushy stunt virus, P19 sequesters ds-siRNA, preventing the degradation of heterologous mRNA transcripts and ensuring their persistence in the cell, thus allowing robust translation and correspondingly high yields of recombinant protein.

The mixed suspension approach is used to achieve coexpression of multicomponent protein complexes; however, its use is limited due to the replicon exclusion phenomenon preventing multiple plasmids with the same origin of replication to productively propagate in the same cell. Coexpression tools bypassing the above phenomenon are vectors harboring multiple copies of a cloning cassette on one plasmid; for example, the pEAQ vector series.

The pEAQ vector series is a derivative of CPMV and direct product of years of research aiming to improve the performance of CPMV as a full virus vector for recombinant expression. CPMV is a segmented sense ssRNA virus. Genome segment 1, RNA1, codes mainly for functions essential for viral replication, whereas the second genome segment, RNA 2, codes for coat protein and a movement protein (Fig. 4.13). RNA-2 replicates in RNA-1-dependent manner. The polyproteins expressed from both segments are cleaved to individual proteins by a protease encoded on RNA-1. Heterologous sequences are inserted as part of RNA-2, while maintaining RNA-1 intact. The discovery of agroinfiltration and the option to provide silencing suppressor *in trans* allowed researchers to replace the genes on the RNA-2 segment with heterologous sequences increasing the size limit of the vector. Mutagenesis in the 5′UTR of the RNA-2 segment removed multiple AUG codons and improved the overall yield of recombinant protein by enhancing translation. The deconstructed viral vector was named CPMV-hypertranslation or CPMV-HT. As a next step, the CPMV-HT cassette, consisting of the 5′ and 3′UTRs of the RNA-2 segment, along with linker allowing cloning of heterologous gene, was incorporated in the backbone of an *E. coli/A. tumefaciens* shuttle/binary vector ensuring productive replication and selection of the plasmid in both hosts, and thus providing a straightforward approach to molecular biology manipulations in *E. coli* and plant (tobacco) delivery via agroinfiltration. Multiple CPMV-HT cassettes can be inserted back-to-back as part of the T-region of the plasmid allowing expression of several polypeptides from a single vector. The pEAQ expression system allows proteins to be harvested within less than a month from the onset of the initial cloning of heterologous sequence and less than 2 weeks after agrofiltration.

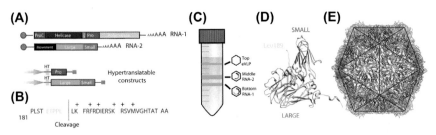

FIGURE 4.13 **Overview of the pEAQ-HT system based on Cowpea mosaic virus (CPMV).** (A) Diagram of CPMV genome (top) and hypertranslatable constructs (*HT*) separately expressing the viral proteinase and the precursor of the L and S subunits (bottom). (B) The sequence of S subunit amino acids 180–213. The C-terminal 24 amino acid segment of S subunit is cleaved following assembly, and is colored magenta. The cleavage site between Leu189 and Leu190 is shown. This region of the polypeptide is highly positively charged and the positive amino acids are indicated. (C) Schematic of CPMV density purification. RNA-1 containing CPMV sediments at the bottom of a Nycodenz gradient (CPMV-B), RNA-2 containing CPMV (CPMV-M) sediment in the middle and empty CPMV particles sediment at the top of the gradient (CPMV-T). CPMV-T particles are the natural equivalent to empty viruslike particles (eVLPs). (D) X-ray crystal structure of the asymmetric unit of CPMV (PDB 1NY7), colored as above. The C-terminal amino acid of the S subunit (leucine 189) is indicated. (E) Model of CPMV capsid (using PDB 1NY7). Each icosahedral particle is comprised of 60 copies of both the L subunit and the S subunit. *Image and figure legend courtesy of Hesketh EL, Meshcheriakova Y, Dent KC, Saxena P, Thompson RF, Cockburn JJ, Lomonossoff GP, Ranson NA. Mechanisms of assembly and genome packaging in an RNA virus revealed by high-resolution cryo-EM. Nat Commun 2015;6:10113.* https://doi.org/10.1038/ncomms10113, *CC-BY,* https://creativecommons.org/licenses/by/4.0/.

The system has been used successfully for expressing and purifying proteins and producing VLPs (Fig. 4.13).

Transient expression although powerful is still in the realm of small scale production as infrastructure for truly industrial volumes is too expensive to maintain. The use of stably transformed transgenic plants continues to be feasible for large-scale manufacturing despite the slower pace of generating plants expressing proteins of interest. Inducible system for expression is being developed based on regulatory elements from bean yellow dwarf geminivirus. Stably inserted construct coding for the protein of interest remains silent until a replication initiation protein is expressed *in trans* from inducible promoter.

Despite being eukaryotes, plants have quite different systems for posttranslational modifications in comparison to mammals. Major differences in glycosylation are raising concerns for possible immunogenicity of human proteins expressed in plants or possible lack of activity due to the somewhat different chemical nature of the synthesized sugars. Efforts to create transgenic plants expressing mammalian glycosylation enzymes and silencing the respective plant systems are in progress.

As more recombinant protein–based medicines are entering the market worldwide and VLP-based nanotechnologies are in development, the importance of highly productive and cost-efficient protein expression systems will continue to rise, welcoming further technological advancements.

Further Reading

1. Brondyk WH. Selecting an appropriate method for expressing a recombinant protein. *Methods Enzymol* 2009;**463**:131–47.
2. Burgyan J, Havelda Z. Viral suppressors of RNA silencing. *Trends Plant Sci* 2011;**16**(5):265–72.
3. Csorba T, Kontra L, Burgyan J. Viral silencing suppressors: tools forged to fine-tune host-pathogen coexistence. *Virology* 2015;**479–480**:85–103.
4. Jarvis DL. Baculovirus-insect cell expression systems. *Methods Enzymol* 2009;**463**:191–222.
5. Kieft JS. Viral IRES RNA structures and ribosome interactions. *Trends Biochem Sci* 2008;**33**(6):274–83.
6. Kneller EL, Rakotondrafara AM, Miller WA. Cap-independent translation of plant viral RNAs. *Virus Res* 2006;**119**(1):63–75.
7. Obembe OO, Popoola JO, Leelavathi S, Reddy SV. Advances in plant molecular farming. *Biotechnol Adv* 2011;**29**(2):210–22.
8. Peyret H, Lomonossoff GP. The pEAQ vector series: the easy and quick way to produce recombinant proteins in plants. *Plant Mol Biol* 2013;**83**(1–2):51–8.
9. Peyret H, Lomonossoff GP. When plant virology met Agrobacterium: the rise of the deconstructed clones. *Plant Biotechnol J* 2015;**13**(8):1121–35.
10. Tschofen M, Knopp D, Hood E, Stoger E. Plant molecular farming: much more than medicines. *Annu Rev Anal Chem (Palo Alto Calif)* 2016;**9**(1):271–94.
11. Zerbs S, Frank AM, Collart FR. Bacterial systems for production of heterologous proteins. *Methods Enzymol* 2009;**463**:149–68.

5

Phage Display

Phage display is a well-established technique for identification, selection, and evolution of protein–ligand interactions widely used to address basic science questions, as well as to develop top-notch technology in diverse fields ranging from research tools to personalized medicine and nanotechnology. Phage display has more than 30 years long history that has reaffirmed the power of the approach and resulted in the development of multiple display techniques to meet the continuous challenge of studying the vast diversity of protein–ligand interactions taking place in living matter. The first reports describing phage display were published in the 1980s and ever since the technique has been continuously improving as advances in molecular biology and various technologies are being used to build on its simple but brilliant principles. Today scientists have in their disposal multiple in vivo and in vitro display techniques allowing sophisticated studies and manipulations of a broad range of protein–ligand interactions. This chapter describes the basic principles of the technique on the example of classical phage display platform based on filamentous phages and provides an overview of current display techniques.

It should be noted that phage display has benefitted from many discoveries resulting from research efforts not focused on the platform directly but rather investigating essential processes, structures, and molecules. As such, the technique provides an excellent example of how gradual science advancements contribute to technology leaps when complexity of living matter is embraced and knowledge is applied critically.

5.1 PRINCIPLES OF PHAGE DISPLAY. BIOPANNING. ADVANTAGES AND LIMITATIONS

Phage display can be simplistically described as an approach to probe protein–ligand interactions between a protein (protein fragment or peptide) "displayed" on the surface of a phage particle and a ligand of interest based on their affinity of binding. If the word phage is omitted from the above sentence, one can easily decide that a big picture definition of affinity chromatography ended up in a wrong paragraph of this work. Indeed, the underlying principle of phage display is that simple. The power of the method stems from the physical connection between the phenotype and genotype of the phage displaying the protein of interest coupled to multiple rounds of selection. In addition, libraries of proteins/ peptides created with a particular goal in mind can be stored long term as DNA clones and probed again for reaching future goals as they arise.

5.1.1 Filamentous Phages as Protein Display Vehicles

Phage display was initially developed using the so-called Ff phages, among which M13 is the most recognizable name due to its multiple applications in molecular biology. In addition to M13, the Ff group includes bacteriophages f1 and fd. All three phages are closely related, sharing 98.5% of their DNA sequence, and have been used for phage display purposes by different research groups (for simplicity, the discussion of filamentous phages will be limited to bacteriophage M13). M13 is a naked virus with circular ssDNA genome and an extended tube-like shape, hence the name filamentous phage. The viral genome is 6.4 kB of length, and codes for 11 genes numbered with roman numbers (I–XI) corresponding to their order. The capsid is composed of the major capsid protein (gp/pVIII), which polymerizes along the length of the genome in several thousand copies, and four minor capsid proteins (gp/pIII, VI, VII, and IX) present in pairs at the tips of the virion in single digit number of copies (Fig. 5.1). Remarkably, the genome can tolerate insertions of foreign DNA up to 12,000 bp, almost twice the length of the wild type.

M13 infects F^+ E. coli strains. The primary viral receptor is found on the F pili on the outer surface of the bacterial cell, whereas the secondary receptor, TolQRA, is part of the inner membrane of the host. M13 attached to the primary receptor is "delivered" to the secondary receptor via a poorly understood sequence of events involving pIII binding, contraction of the pili, and virion passage through the outer membrane of its Gram-negative host. The virion binding to the secondary receptor results in phage uncoating at the inner E. coli membrane and transfer of the viral genome inside the host cell. M13 replicates as an episome in the cytoplasm of the host using circular dsDNA replicative form and a rolling circle mechanism. As more M13 replicative forms are produced and more phage proteins are synthesized, critical pV concentration is reached allowing the protein to bind to the genome packing sequence and initiate virion assembly. The genome DNA is initially tethered to an inner membrane complex of viral and host proteins, which is instrumental for passing the newly made genomes into the periplasmic space. All capsid proteins are synthesized as integral inner membrane proteins and their insertion is guided by leader sequences. Virion assembly takes place in

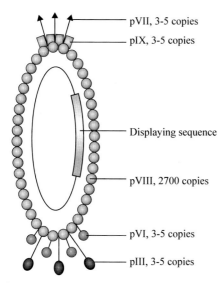

pVII, 3-5 copies

pIX, 3-5 copies

Displaying sequence

pVIII, 2700 copies

pVI, 3-5 copies

pIII, 3-5 copies

FIGURE 5.1 **Structure of the bacteriophage M13.** Bacteriophage M13 is a Class II ssDNA virus infecting *E. coli*. The wild-type particle is approximately 900 nm long tube/filament whose walls are built from 2700 molecules of the major capsid protein, PVIII. The tips of the capsid tube are "decorated" with minor capsid proteins, PIII, PVI, PVII, and PIX, all of which can be utilized in low-density surface displays. The size of bacteriophage M13 virion can vary significantly as a function of the length of packaged DNA.

the periplasmic space between the inner and outer host membranes with the assistance of several specialized assembly proteins that are not packed into the viral particle. Mature virions are released without cell lysis. The latter, together with the lack of replication regulation and the ability of the episome to be transferred to daughter cells during cellular division, allows production of high titer viral stocks (~10^{13}), which is a major factor for the success of the M13 phage display platform. The dsDNA replicative form can be manipulated as a plasmid allowing for straightforward insertion and/or mutagenesis of a protein/peptide of interest. The various copy numbers and properties of the capsid proteins allow multiple options for configuration of the displayed protein/peptide (Fig. 5.2). High density and polyvalent displays increase the chances of selecting ligands; however, they are productive only for small size proteins and peptides due to steric interference. Low density and monovalent displays are preferable when larger proteins are to be displayed. The density/valency of display can be varied by coexpressing both wild-type and peptide/protein fusion version of the same coat protein. Depending on the structural and functional constraints on the displayed protein/peptide, options for N- and C-terminal fusions are available.

5.1.2 Key Steps in Phage Display Technology

Phage display is a multistep process (Fig. 5.3) that starts with library design. Libraries contain zillions of DNA clones harboring sequences of interest, each individually cloned in the same DNA backbone with

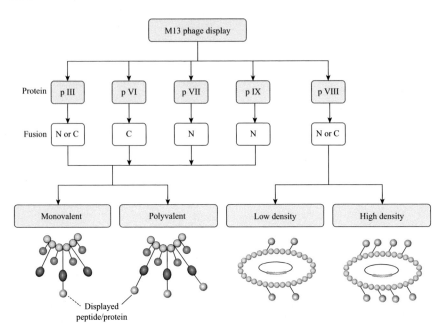

FIGURE 5.2 **M13-based phage display alternatives.** All M13 capsid proteins have been explored as possible display platforms. They offer various options for N-, C-, or both termini fusions, as well as opportunities for different densities/valences of display. High-density display is achieved by using the major capsid protein as a display vehicle for peptides and smaller size proteins. Display of larger size proteins could be accomplished by low-density display using PVIII. The density of the display is controlled by the availability of wild-type PVIII proteins during virion assembly. Similarly, display on any of the minor capsid proteins can be executed in monovalent or polyvalent modes depending on the presence or absence of wild-type version of the display vehicle.

multiple regulatory sequences allowing effective replication, transcription, translation, and display of the corresponding proteins/peptides. A classical phage display experiment, where a library of different variants of the same protein is screened to identify high affinity binders to the ligand of interest, starts with cloning the protein in question in a phagemid vector or a recombinant phage and demonstrating that it is expressed in a functional form on the surface of the produced phage particles. Standard practices involve incorporation of a protein tag that is easily detectable by ELISA and/or Western blot and potentially can be used as a tag for affinity purification of the protein in downstream steps of the protocol. Selection of tags is driven by their compatibility with the protein under investigation, as well as by the accessibility of reagents for detection. Traditional protein tags employed in protein expression, purification, and detection, such as FLAG tag, HA tag, His tag, etc. are commonly utilized in phage display. Phagemid vectors are the preferred tool compared to phage vectors due

1. Library construction

Target

2. Library screening

Biopanning

Unbound phage

Bound phage

3-5 cycles of enrichment

Phage elution

3. Clone isolation

Individual plaques

4. Amplification in *E.coli*

5. Analysis

| Specificity validation | Affinity measurement | DNA sequencing |

Clone identification

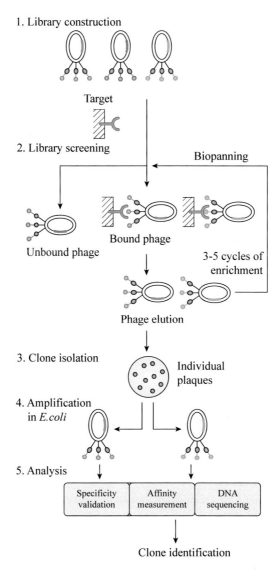

FIGURE 5.3 **Key steps in phage display technology.** Phage display libraries are created by conventional cloning techniques using the dsDNA replication form of the phage or phagemid genome. Libraries are transformed in competent *E. coli* cells and recombinant phages produced. The phage pool is then screened against a target of interest and phages binding to the target are identified. Binders with higher affinity are identified by multiple rounds of selection. The pool of the eluted high-affinity binders is propagated in *E. coli*, individual clones isolated and characterized. Further analyses are performed to confirm the specificity of the selected phages, to measure the affinity of the binding, and to identify the sequence of the selected clones.

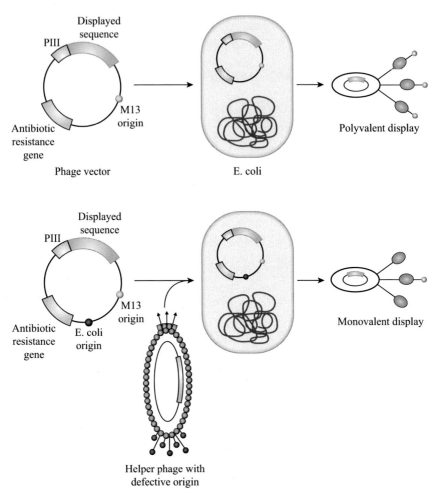

FIGURE 5.4 **Properties of phage and phagemid vectors.** M13 phage display can be executed using phage or phagemid vectors. Phage vectors contain an M13 origin of replication, all essential genes for the phage propagation along with sequences coding for the protein/peptide to be displayed. Phagemid vectors contain origins of replication from M13 and *E. coli* and code for a fusion of the protein/peptide to be displayed and a capsid protein being used as a display vehicle. Genes essential for phage packaging are provided in *trans* from a helper M13 phage with a defective origin of replication. Packaged virions contain the phagemid genome, allowing for easy identification of the DNA sequence of selected proteins/peptides.

to their stability and ease of manipulations (Fig. 5.4). Phagemid vectors contain plasmid and phage origins of replication, DNA packing sequence, selective marker, and a coat protein cassette supporting the cloning, the expression and detection of the protein/peptide under investigation. Phagemids can be replicated independently in the host cell as regular plasmids or by phage machinery when a helper M13 phage infects the cell harboring them. The first mode of propagation allows phagemids to be manipulated with molecular biology tools, whereas the second mode allows them to be packaged as phage particles displaying the protein/peptide under investigation. The mere expression of the protein of interest on the surface of a phage particle does not guarantee proper folding and functionality, a problem commonly encountered in heterologous protein expression that can be addressed more or less empirically on different levels. Therefore the availability of an assay confirming the functionality of the expressed molecule is a huge stepping stone toward the success of phage display. Once the expression of functional protein under investigation is confirmed, a library of multiple variants is created using conventional methods for mutagenesis. The specific mutagenesis approach varies depending on the goals of the experiment and could involve use of degenerate oligonucleotides, error-prone PCR (polymerase chain reaction), Kunkel mutagenesis, bacterial strains with error-prone replication, etc. To ensure meaningful screening, the library size has to be 100-fold larger than the theoretical number of possible variants. In addition to protein and peptide libraries, phage display can be performed with cDNA libraries. The most sophisticated screen of cDNA libraries is known as ORF (open reading frame) phage display in which cDNA libraries are specifically designed to exclude noise-creating non-ORF clones.

Library selection screening (biopanning) is accomplished by exposing the ligand of interest to the generated library of protein variants displayed on the surface of phage particles. There are multiple possibilities for technical execution of the screening process. Screening protocols using immobilized ligands are performed in the context of columns or microtiter plates, whose nonspecific binding surfaces have been blocked to reduce background. Screening schemes with soluble ligands use biotinylated molecules, which bind to cognate phage particles in solution and are then "collected" on streptavidin-covered surfaces. Cell surface ligands can be screened in the context of cell monolayer, cell suspension, tissue preparations, or in whole animals. The latter involves removal and analysis of tissue sample following phage library injection. Phage particles selected in the biopanning process are recovered by breaking the protein/ligand interactions with conventional methods such as elution with salt, detergent, changes in pH, etc. The stability of bacteriophage M13 in such conditions is a major factor for the success of phage display as an experimental approach. Eluted clones are propagated in *E. coli*, the specificity and affinity of their binding confirmed and further analyzed as needed depending on the end goal of the experiment. Finally, the DNA of the eluted particles is recovered and sequenced, thus identifying the binding partner of the researched ligand on the level of the genotype and enabling further modifications of the displayed protein/peptide. Premade phage display libraries of random peptides with various lengths are commercially available from several companies and can be utilized to identify binding partners of any ligand of interest. Since M13 capsid is built from multiple capsid proteins, it allows for multifunctional phage display in which more than one capsid protein is fused to heterologous protein/peptide. This feature is especially valuable to applications relevant to nanotechnology and nanodevices.

5.1.3 M13 Phage Display Limitations, Modifications, and Extensions

Despite many successful applications, M13-based phage display has its limitations stemming from the biology of the bacteriophage M13 itself. All capsid proteins used in phage display are essential for phage growth, thus their manipulation could be rather tricky limiting the size of the protein segments fused to them, as well as the type of the fusion. For example, pVI can tolerate fusions only at its C-terminal end, thus not allowing proteins that need their N-terminus free for function to be displayed in the context of pVI. The copy number of the major capsid protein offers great opportunity in terms of display density; however, that limits the size of the displayed molecules since it introduces space constraints for displaying large molecules. On the other hand, the low copy number of the minor capsid proteins may present an obstacle for the selection of low-affinity binders. While the lysogenic life cycle of M13 is a huge advantage for the system permitting very high viral titers, the virus assembly in the periplasm may present an obstacle for the displayed proteins to get across the inner host membrane, especially if they are highly soluble. Similarly, while the option to insert long chunks of foreign DNA and package them could be a great advantage, the resulting phages are bigger/longer, and thus less stable and with a slower rate of propagation, both of which can easily translate into disadvantages during the screening process. The described limitations have driven extensions and modifications of the classical M13 phage display ultimately resulting in developing bacteriophage, cell, and in vitro alternatives for display vehicles to enable the application of the principles of the technique to molecules for which the classical version is not the best match.

5.1.4 Phage Display Alternatives of M13

Phage display systems based on bacteriophages T7, T4, and Lambda have been successfully developed and used for various purposes. Table 5.1 summarizes key characteristics of the above phage particles and their advantages as phage display vehicles. Generally, they extend the capabilities of the technique by (1) allowing larger proteins to be displayed due to their capabilities to carry foreign DNA with larger sizes;

TABLE 5.1   Properties of Bacteriophages Utilized in Phage Display

	M13	**T4**	**T7**	**Lambda**
Genome size	6.4 kB	168 kB	40 kB	48.5 kB
Display protein	All coat proteins, usually pIII and pVIII	SOC and HOC	gp10	pD, pV
Display limit	110 kDa on pIII 10 kDa on pVIII	710 kDa	132 kDa	600 kDa
Protein function	Essential	Nonessential	Essential	Essential
Display density	3–5 copies on pIII Up to 2700 copies on pVIII	Up to 810 copies on SOC Up to 155 copies on HOC	Up to 415 copies	Up to 405 copies on pD 6 copies on pV
Life cycle	Lysogenic	Lytic	Lytic	Lysogenic

(2) increase the success rates of displaying soluble proteins since they are fully assembled in the cytoplasm and leave the host cell by lysis, thus the displayed soluble proteins do not need to cross the cellular membrane; (3) offer possibilities for various densities of display. In addition, bacteriophage T4 supports protein display using fusion to nonessential capsid proteins (SOC and HOC), thus minimizing potential detrimental effects of the fusion on particle stability and preventing possible screening bias. The Lambda system allows for multivalent display since gpD, the major capsid protein, is in fact a trimer. The diversity of phage display vehicles ultimately brings the question, which system can be connected to best experimental outcomes. As in many other similar cases, the best answer to such questions is "it depends." Logically, the better the match between the strengths of the particular vehicle and the characteristics of the molecules involved, the bigger the chances of straightforward selection process are. One study surveyed the autoantibody profiles of cancer patients using T7 and Lambda phage display and found that for some antigens both approaches delivered comparable results, whereas other groups of antigens were preferentially identified by the T7 or Lambda system.

Eukaryotic viruses, such as tobacco mosaic virus, baculoviruses, adenovirus, are also being used to display molecules of interest utilizing the principles of phage display, and are currently considered valuable vehicles in development of new vaccines, as well as possible resources in development of virus-based nanomaterials. They have the major advantage of tolerating large insertions of heterologous genetic material, as well as being able to display proteins that need posttranslational modifications for their functionalities.

5.2 THE DISPLAY TECHNIQUES FAMILY

The success of the phage display technique and the diversity of protein/peptide–ligand interactions in living matter have triggered the development of a variety of display techniques further expanding the power of the approach. Cell surface display systems have been described in bacteria (both in Gram-positive and Gram-negative species) and yeast. Cell-based display systems work in a similar manner to phage display: peptides and proteins are expressed on their surface and utilized in screening processes to identify the interaction/ligand of interest; however, in contrast to phage systems, they are autonomously replicating and have multiple advantages due to their size. Larger size offers larger display surface and larger number of displayed molecules, which is an advantage for screening of weak protein/ligand interactions. Both bacterial and yeast cells can be monitored with optical methods, thus allowing straightforward quantitation of their growth and density, as well as selective labeling and sorting of cells of interest, assuming that relevant fluorescent ligands are available and FACS can be used for detection. In addition, yeast display systems can account for the role of posttranslational modification in eukaryotic proteins. Similar to phage display, cell surface display techniques are limited by the transformation efficiencies of the respective cells (10^{9-10} for prokaryotic systems and 10^7 for yeast display), the displayed molecules are expressed in vivo, and the screening process takes place in vitro. The linkage between genotype and phenotype can be established easily by sequencing the relevant DNA coding elements, assuming cells are not lysed.

In vitro display techniques (ribosome display, cDNA/mRNA display, CAD display, CIS display, SNAP display, and bead surface display) capitalize on in vitro expression and selection and allow much larger libraries (10^{13-14} clones) of proteins/peptides to be screened compared to the in vivo display techniques since transformation efficiency is no longer an issue. The key features of in vitro display techniques are summarized in Table 5.2.

The plethora of display techniques developed in the last several decades combined with technological advancements of next generation sequencing, mass spectrometry, and extended computational modeling capabilities have made display technologies the gold standard in selecting molecules with properties of interest and evolving them for the purpose of practical applications in many different areas of science.

5.3 PHAGE-ASSISTED CONTINUOUS EVOLUTION APPROACH

Whereas the numerous variations of display techniques summarized above significantly broaden the applicability of the original phage display technique and address various limitations, the phage-assisted continuous

TABLE 5.2   Overview of Display Technologies

Technique		Display Principle	Selection Principle	Library Diversity
Cell-based	Phage display	Protein/peptide library is displayed on the surface of phage particles as fusion to coat protein	Capture/ elution	10^{9-10}
	Cell surface display	Protein/peptide library is displayed on the surface of a living cell as fusion to cell surface protein	FACS selection employing fluorescent ligand	10^7 yeast 10^{8-10} E. coli
In vitro	Ribosome display	Protein/peptide library is displayed on the surface of stalled ribosome/mRNA (lacking stop codon) complex	Capture/ elution	10^{13-14}
	cDNA/ mRNA display	Protein/peptide library is displayed as covalently attached peptides to cognate mRNA/cDNA hybrid	Capture/ elution	10^{13}
	CAD and CIS display	Displayed proteins/peptides are covalently attached to cognate DNA via *cis*-acting DNA-binding protein (RepA in the CIS technology and P2A endonuclease for the CAD technology)	Capture/ elution	10^{12}
	SNAP tag display	DNA labeled with benzylguanidine (BG) is packed into microemulsion droplets (1 clone per droplet) and in vitro expressed into a SNAP tag-fused peptide/ protein. The SNAP tag and its substrate BG covalently interact creating a physical linkage between DNA and corresponding peptide/ protein.	Capture/ elution	10^{10}

evolution (PACE) approach takes its conceptual principles to a new level by combining them with continuous host culturing and introducing high level of automatization, thus allowing molecules to be evolved over the course of days with reduced efforts for direct hands-on manipulations. The first proof of principle work demonstrated that the specificity of T7 RNA polymerase can be changed to allow the enzyme to work using the T3 promoter over the course of 8 days with minimal human involvement. The PACE principle takes advantage of the indispensable nature of the pIII capsid protein of bacteriophage M13, broad molecular biology knowledge allowing for controlled mutagenesis via the expression of an error-prone polymerase, and continuous flow culturing of the *E. coli* in a mode limiting the impact of mutagenesis on host genes. If one focuses on the macroscopic level, PACE can be described as a brilliantly designed process of growing phage in a single flask, called a lagoon, equipped to receive continuous flow of host *E. coli* cells and key chemicals while allowing waste removal (Fig. 5.5). PACE can be applied to evolving any molecule of interest that can be coupled to transcription. The host cells are harboring accessory plasmid (AP) coding for pIII and mutagenesis plasmid coding for an inducible error-prone polymerase. The host cells are infected with an engineered M13 phage in which the gene for pIII is replaced with the gene of interest, in the discussed case T7 RNA polymerase. Since pIII is an essential gene, the engineered M13 phage will propagate only if pIII is provided *in trans*, by the AP present in the host cell. The gene for pIII is under the control of a bacteriophage T3 promoter and will be expressed only when the T7 polymerase is able to initiate its transcription, i.e., when T7 polymerase acquires the ability to recognize the T3 promoter. The synthesis of error-prone DNA polymerase is induced by adding inducer molecule (arabinose since the *ara* operon promoter is used to control the expression) in the culturing flask and the M13 life cycle unfolds with its natural rate. Since the engineered M13 phage is fully capable of replication, many genome copies will be produced and coded proteins expressed, including the protein subject to evolution. The error-prone DNA polymerase introduces many mutations, some of which will ultimately confer change of specificity of the T7 RNA polymerase. Thus some pIII protein will be expressed allowing packaging of M13 progeny, ready to infect more host cells. The continuous flow of fresh bacterial culture ensures that all host cells leave the lagoon before they have a chance to divide, thus eliminating accumulation of mutations in the host genome. At the same time, the flow is slow enough to allow the production and accumulation of infectious phage particles carrying T7 polymerase mutants with increased specificity for the T3 promoter. The latter is possible due to a much faster rate of phage replication as compared to the rate of host cell division. Over time, continuous culturing and mutagenesis results in a population of phages encoding T7 polymerase, with continually improving specificity for the T3 promoter. Culture media is periodically withdrawn from the system and plated to identify and quantify resulting phages. The evolved protein is tested for the target activity and the corresponding phage DNA sequenced to identify the underlying mutations. The outcomes of the PACE proof of principle work can be summarized as demonstration of fast-forward in vivo process in which a double-digit number of evolution rounds can be accomplished in a single day without any human involvement. As with any new experimental approach, one has to ask about the potential of success when the approach is applied to a random system of interest. PACE works only for systems that can be connected somehow to transcription, which is a drawback; however, on the grand scale of things, if there are to be limitations

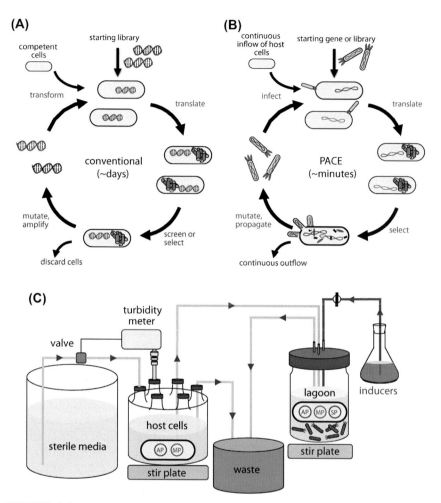

FIGURE 5.5 **Phage-assisted continuous evolution overview.** Directed evolution cycles in (A) conventional-directed evolution in cells and (B) phage-assisted continuous evolution (PACE). Steps in the evolution cycle that typically require the intervention of the researcher are shown in red; those that do not are shown in green. (C) Schematic of the PACE apparatus. Host *E. coli* cells maintained at constant cell density continuously flow through a lagoon vessel (along with optional chemical inducers) containing phage at dilution rates of ~1.0–3.2 lagoon volumes per hour. Host cells harbor and accessory plasmid (AP) coding for M13 pIII protein and mutagenesis plasmid (MP) coding for DNA polymerase dominant negative proofreading subunit dnaQ926 and the error-prone repair DNA polymerase pol V, which suppress proofreading and enhance lesion bypass, respectively. Collectively, the induction of both polymerases increases the cellular mutation rate ~100-fold. The selection plasmid codes for the protein subject to protein evolution, which in the case of the proof of principle study codes for a T7 RNA polymerase and it is introduced in the host cell via M13 phage infection. Only mutant RNAPs with changed specificity can initiate transcription of the pIII gene (placed under the control of heterologous promoter) and support phage propagation. Multiple rounds of evolution eventually result in selection of an enzyme with desired properties. *Image and parts of the figure legends are reprinted from Esvelt KM, Carlson JC, Liu DR. A system for the continuous directed evolution of biomolecules.* Nature *2011;**472**(7344):499–503. With permission from Nature publishing group.*

(and there are always some, otherwise one will be using magic and not science), being limited to transcription is not that bad simply due to the vast knowledge of the process, especially in *E. coli*, and the availability of many molecular biology tools in the field. Preexisting detailed knowledge can contribute significantly to the success of PACE by providing "stepping stones" to the selection process. For example, when a native T3 promoter was used to drive the pIII expression from the AP in the above described experiment, no meaningful results were observed; however, when a hybrid T7/T3 promoter was used, the selection process worked fast and efficiently, allowing to identify a T7 polymerase mutant with several hundred-fold increase in affinity for the T3 promoter. There was only one nucleotide difference between the T3 and the T7/T3 hybrid promoter. Since the initial report of the PACE principle, several works have focused on various extensions of the techniques and it has been demonstrated how PACE can be used to assess potential for development of drug resistance against antiviral drugs on the example of inhibitors of Hepatitis C protease. Undoubtedly, more PACE developments and applications are to come.

5.4 APPLICATIONS OF PHAGE DISPLAY

Since protein/ligand interactions are involved in essentially every process related to living matter, the applications of phage display are rather numerous and diverse. Technically, any process, material, product, etc. that uses a molecule selected by phage display can be viewed as application of the technique. On a big picture level, applications fall into three major categories: (1) utilization of proteins selected by phage display; for example, antibodies and enzymes with improved/novel properties; (2) utilization of peptides selected by phage display to functionalize other entities; for example, liposomes or gold particles for the purpose of chemotherapy or photothermotherapy; and (3) utilization of whole phage particles with properties of interest; for example, for development of sensors for pathogenic bacteria or scaffolds for tissue regeneration and engineering. Several examples of each group will be briefly discussed with emphasis on applications related to biomedicine and biotechnology. Examples of nanotechnology applications of phage display are discussed in Chapter 6.

5.4.1 Protein-Based Applications of Phage Display

The therapeutic and research applications of antibodies are invaluable. Phage display allowed scientists to select and evolve antibodies in vitro in part replacing conventional immunization and hybridoma techniques. Antibodies (described on the example of IgG, the antibody type conferring long-lasting immunity in the human body, Fig. 5.6) are very complex large molecules

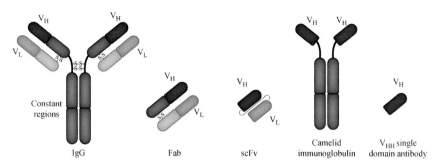

FIGURE 5.6   **Antibody fragments manipulated by phage display.** Antibodies are complex molecules with a sophisticated quaternary structure stabilized by disulfide bonds (far left). Their complexity does not allow for straightforward direct selection via phage display. Instead Fab fragments (second left) are used in phage display experiments. Fab fragments are composed from the variable regions of the heavy and light chains (V_H and V_L) and can be viewed as a monovalent antibody. An alternative version of Fab fragments is the single-chain Fv antibody fragments, in which V_H and V_L chains are connected via a long linker. Although significantly less complex than conventional antibodies Fab and scFv fragments are challenging to produce since they require proper domain association. The discovery of heavy-chain only camelid antibodies (on the right) allowed for simplified antibody selection and production. Usually a single V_H domain library is created and high-affinity binders are selected via biopanning. Standalone V_H molecules are used as V_{HH} antibodies or heavy-chain only antibodies are generated via recombinant DNA technology.

(~150 kDa) built of four polypeptide chains, commonly described as heavy (in gray) and light (in orange) chains. Each chain is built of constant (C_H in light gray and C_L in light orange) and variable regions (V_H and V_L in darker gray and orange, respectively). The four chains are associated with each other via several sets of disulfide bonds. Each IgG molecule harbors two antigen-binding sites formed by the variable regions. Each variable region has three major specificity determinants known as complementarity-determining regions or CDRs (not depicted in Fig. 5.6 for simplicity). Full-range antigen recognition and specificity is retained by the so-called Fab fragment (~50 kDa), initially identified in antibody proteolysis studies. The Fab fragment is composed of the full-length variable regions and one additional constant region for each chain. The Fab fragments are less stable than the full-size antibodies; however, that issue has been circumvented by inventing the scFv (single chain variable fragment) antibody fragment comprised only of the heavy and light variable regions connected with a flexible linker. While the scF_V fragments retain the antigen-binding properties of a full-size antibody molecule, their size is only ~25 kDa and their complexity is significantly reduced, thus removing the major hurdles for antibody production in vitro. Initially, antibody phage display libraries were generated using naïve or "immune" B-lymphocytes. The sequences of the variable regions of the heavy and light chains were amplified by RT-PCR and cloned on phagemid vectors connected with a linker (Fig. 5.7). Since the PCR and the cloning were done in bulk, phagemid clones

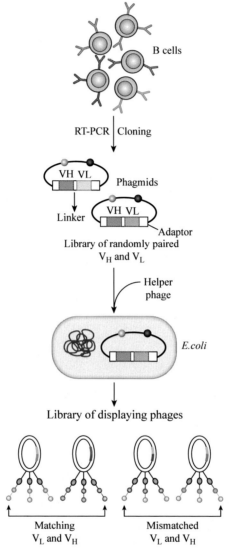

FIGURE 5.7   **Generation of antibody phage display library.** The genes for the variable heavy and light chains are obtained by RT-PCR from populations of B lymphocytes. The pool of V_L genes and the pool of V_H genes are ligated together on a single phagemid to create a library of V_L/V_H fragments. Each phagemid from the library contains a random combination of fragments, including many mismatched pairs. The library is packaged into M13 phages displaying different combinations with the assistance of a helper phage and subjected to biopanning.

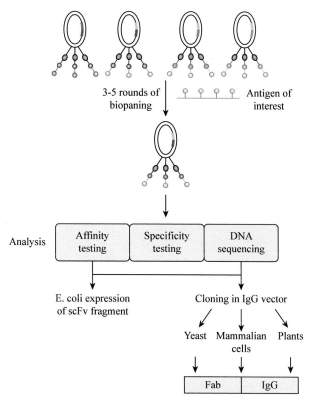

FIGURE 5.8 **Biopanning of antibody phage display library.** Antibody phage display library is subjected to 3–5 rounds of biopanning against the antigen of interest. Individual selected binders are isolated and propagated to allow detailed analysis of their specificity and affinity of binding. Once selected clones are sequenced, the identified DNA fragments are cloned in different vectors to allow production of scFv fragments in *E. coli* or Fab/IgG molecules in yeast, mammalian tissue culture or plants.

with random combinations of V_H and V_L fragments were generated. Each library contained a mix of precisely matched fragments and mismatched fragments. Biopanning against the antigen of interest identified a pool of specific binders (Fig. 5.8; only one binder is shown for simplicity), which was enriched in high-affinity ones via multiple rounds of selection. The selected phages were then propagated in *E. coli* and further analyzed to confirm their specificity and measure their affinity. The DNA fragments corresponding to the displayed variable regions are sequenced and the DNA sequences applied forward to different downstream protocols depending on the end goal. scFv fragments are expressed in *E. coli*, whereas Fab and IgG molecules could be produced in yeast, mammalian tissue culture, or plants.

A variety of approaches have been applied toward generation of antibody libraries. Table 5.3 lists the characteristics of major types of libraries and their applications. Each type of library offers unique advantages

TABLE 5.3   Types of Antibody Phage Display Libraries

Library Type	Displaying Sequence	Bias	Antibody Affinity	Targeted Application
Naive	Genes derived from naïve B lymphocytes of healthy nonimmunized individuals	Biased against self-antigens	Variable	Diagnostics
Immune	Genes derived from B lymphocytes undergone clonal selection and affinity maturation as a consequence of immunization with antigen of interest	Antigen-biased	Very high	Studies of infectious disease, research tools, vaccine development
Synthetic	Randomized V-genes assembled by polymerase chain reaction	None	Variable	Selection of antibodies outside natural repertoire; research tools
Semi-synthetic	Naturally selected V-genes with randomized CDR(s)	Antigen-biased	Variable	Research tools, improvement of natural antibodies

and most importantly, once created, libraries can be stored long-term allowing biopanning to be performed against multiple targets over time and/or already selected antibodies to be further evolved. A major breakthrough in the field was the discovery of the camelid antibodies (Fig. 5.6) and the realization that single chain variable domains are capable of antigen recognition and binding. Camelid antibodies are named after the family Camelidae (camels, llamas, alpacas), from where they were isolated. They are composed of heavy chains only and lack light chains, thus accomplishing antigen binding solely by the heavy chain variable region (V_{HH}). The designation V_{HH} was introduced to account for the differences between the heavy chain variable region in camelids and humans. Although both of them contain three CDR regions, their general consensus is somewhat different. In addition, key hydrophobic amino acids involved in binding the V_L region in human antibodies are replaced with hydrophilic residues conferring V_{HH} stability and solubility. Individually expressed, V_H regions of human and mouse antibodies capable of antigen-binding, however, tend to be unstable and to easily aggregate due to the absence of the V_L interacting partner shielding the hydrophobic amino acids from the water-based environment. "Camelization" of human V_H fragments and phage display screens with the end goal to increase stability, allowed for the development of single-domain human antibody fragments. Alternative antibody fragments such as diabody, minibody, etc. have been engineered and are described in the specialized literature. The concept of single-domain antibodies contributed greatly to advancement of the field. Single variable domains are compact (~15 kDa), much easier to mutagenize, as well as to express productively and produce in large amounts. Synthetic phage display libraries offer unparalleled opportunities for antibody selection without any bias and streamlined approaches to antibody maturation and evolution. The first big success story of a drug designed with phage display came on the example of Humira (human monoclonal antibody in rheumatoid arthritis), a monoclonal antibody against TNF-α (Tumor Necrosis Factor α), which modulates the inflammatory response in autoimmune diseases, such as rheumatoid arthritis, psoriasis, and Crohn's disease, among others. Humira targets free TNF-α, thus preventing the molecule from binding to its receptor and thus resulting in inflammation suppression. According to the Genetic Engineering and Biotechnology News website (http://www.genengnews.com), Humira was the topselling drug generating a little over 16 billion US dollars in sales, up to 14% compared to the previous year. The worldwide patents of the drug are starting to expire and generic low-cost versions are starting to appear on the market.

Applications of antibodies as drugs, diagnostic, and research tools are rather extensive and their availability has revolutionized many fields. Creative approaches have been developed recently to take advantage of the single-domain and camelid antibodies, in the realms of new research tools design and biomedical research. For example, V_{HH} antibody fragments have been used to develop new reagents for live microscopy and indirect immunostaining, both widely used in biomedical research and immunochemical diagnostics (Fig. 5.9). V_{HH} domains fused to green fluorescent proteins (GFP in the depicted example) are used as a probe in live microscopy, tracking the location of an antigen of interest by following the green color of GFP. The fusion is inserted in cells using conventional transfection or transduction with virus-based vectors. The small size and high specificity of the single domain antibody minimize the chances for interference with the function of the tracked proteins. Phage display capabilities to select antibodies against alternative conformations and the availability of alternative fluorescent proteins allow for detailed mechanistic studies in vivo on the single cell level. Antibodies against different conformations/transition states of the same protein have been selected by biopanning against the corresponding form of the protein stabilized by covalent modifications or changes in oxidation/reduction state. Indirect immunostaining relies on the coordinated action of a specific antibody recognizing the antigen of interest (primary antibody) and fluorescently labeled/enzyme-conjugated secondary antibody. Secondary antibodies

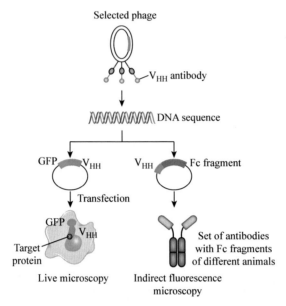

FIGURE 5.9 **Examples of phage display applications for development of microscopy research tools.** DNA derived from the V_{HH}-displaying phage selected by biopanning is sequenced and subsequently cloned as a fusion to green fluorescent protein GFP (left) or as a fusion to constant regions of camelid-like antibodies (right). The V_{HH}–GFP fusion is used for tracking protein of interest in live microscopy experiments utilizing the specificity of the selected V_{HH} antibody (recognizing a protein from the nuclear membrane in this example) and the fluorescence capabilities of GFP. Sets of camelid-like antibodies, containing Fc fragments from different animals are created for the purpose of indirect immunofluorescence imaging. The approach relies on the specificity of the selected V_{HH} antibody and the fluorescence of secondary anti-Fc fragment antibodies conjugated to a fluorescent dye.

are raised against the Fc fragment of particular animals, allowing the same secondary antibody to be used with multiple primary antibodies. Often, when multi-antigen immunostainings are performed, it is challenging to assemble a working combination of primary and secondary antibodies to avoid cross-reactivity. Since camelid antibodies are smaller and less complex, their manufacturing is more straightforward. Genetic engineering approaches swapping variable and constant regions allow sets of antibodies with identical variable domain and alternative Fc domains (for example, mouse, rabbit, goat, etc.) to be easily created, thus allowing a mini reagent library to be assembled with enough choices to satisfy any primary/secondary antibody pairing needed. In clinical research, single-domain antibody fragments are being tested as a tool to combat viral infections and cancer. Proof of principle experiments have shown that expression of antibody fragment (scFv) against LANA 1 (latency associated nuclear antigen 1) of Kaposi's Sarcoma-associated herpesvirus (KSHV) showed promising

TABLE 5.4   Phage Display Targets for In Vitro Biopanning and Relevant Applications

Target	Outcomes	Projected Applications
Receptors	Selection of receptor agonist(s), antagonist(s), and antibodies	Research advancement, drug development
Antibody	Epitope mapping, specificity studies	Vaccine development, antibody-based drugs development, research advancement
Antigen	Antibody selection and affinity improvement	Research advancement, antibody-based drugs development
Enzyme	Understanding of substrate specificity, inhibitor selection	Research advancement, drug development
Any protein of interest	Identification of interaction partners	Research advancement, drug development
Serum samples	Selection of disease-specific antibodies, identification of autoantibodies and autoantigens	Vaccine development, developing drugs/treatments for autoimmune diseases

results in tissue culture as a potential strategy to neutralize KSHV latency. The antibody interfered with the maintenance of plasmids containing the LANA1-binding sequences and artificial chromosome harboring the entire KSHV genome. Similarly, V_{HH} antibody fragments against Hepatitis B antigen caused up to a two-fold reduction in the titer of produced viruses in a mouse model. Intracellular expression of antibodies against the vascular endothelial growth factor receptor (VEGF R2) reduces tumor growth in the human melanoma xenograft model. Antibodies against plant pathogens expressed in plants confer resistance to the antigen (bacteria, virus, nematode, or fungus). In all described cases, the expressed antibody fragments were selected by phage display and delivered intracellularly via gene therapy vectors or conventional plant transformation approaches.

Although phage display for protein selection is best known as a tool for antibody selection and maturation, the technique is very useful when applied to any protein of interest. Functional classes of proteins most frequently targeted in in vitro biopanning experiments are listed in Table 5.4 along with relevant applications. In addition to in vitro biopanning, phage display libraries can be screened in vivo in animal models or against cells in tissue culture settings. In the interest of space, these modes of biopanning will be discussed in the next two subsections; however, they are fully relevant to protein-displaying phage libraries.

Today's science has at its disposal extensive genome sequencing data, and the whole genomes of a diverse group of organisms, including human, have been sequenced. A long-term goal in the realm of genomics and proteomics is to assemble a collection of antibodies against all human proteins and to use them to create tissue distribution maps. A combination of phage display and yeast display has been used to attempt assembly of pairs of human antigens and antibodies binding to them. One library displays the antigens and the other antibodies. Both libraries are mixed resulting in retention of binder phages to the surface of cognate yeast cells. Nonbinder phages are washed away. Next, the bound phages are visualized by fluorescent labeling and yeast cells–fluorescent phage pairs sorted out by FACS. Following phage elution, both yeast and phages are amplified independently and their DNA sequenced to identify the individual clones in the analyzed pair. The approach was found to be technically challenging; however, the library versus library biopanning idea has the potential to yield useful outcomes that are not reachable when experimenting in one protein at a time mode.

5.4.2 Peptide-Based Applications of Phage Display

Just like protein molecules, peptides can be either displayed or used as a target to screen against in phage display experiments. Both approaches have selected many useful molecules with applications in diverse fields. The next example is an illustration of the principle of biopanning against live cells demonstrating the breadth of the phage display technique (Fig. 5.10). Targeted delivery of drugs in the body is a huge challenge for medicine. That is especially true for chemotherapy drugs that are expected to execute their effect based on differences in the rate of DNA metabolism and cellular division between cancer cells and healthy cells. Targeted delivery allows high dose of chemotherapeutic agents to be delivered to tumors directly and specifically, thus minimizing the side effects on healthy cells and the chances for new cancerous growth to arise later due to DNA damage inflicted by the chemotherapeutic treatment itself. In principle, targeted delivery is achievable if knowledge for the specific molecular landscape of cancer/diseased cells is available along with a vehicle capable to deliver the necessary drugs in a bioactive form. Parallel biopanning of healthy and cancer/diseased cells allows molecules specific for the diseased cells to be identified along with the peptides binding to them. The identified peptide can then be used to modify the surface of medicine delivery vehicles, thus providing specific destination address and achieving selective delivery. Peptides specific for cancer cells have been incorporated on the surface of liposomes, viruses, and virus-like particles, which offer robust shells with good loading capacity and many possibilities for encapsulation of small molecules, protein-based drugs, or nucleic acids coding for such. In addition, the strategy is applicable to the delivery of molecules capable of genome

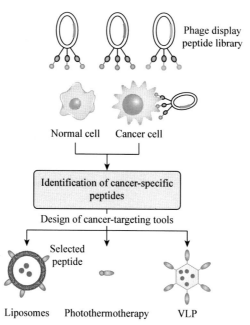

FIGURE 5.10 **Examples of phage display applications for development of anticancer therapeutic approaches.** Cancer cells-specific peptides are selected by phage display and incorporated on the outer surface of liposomes, virus-like particles (VLPs) or gold nanoparticles, directing the delivery of the attached structure specifically to cancer cells. The internal cavity of liposomes and VLPs are loaded with anticancer drugs, which are delivered in high doses only to cancer cells ensuring maximum therapeutic effect while minimizing damage to healthy cells. Cancer cells "decorated" with gold nanoparticles are irradiated with laser beams, resulting in absorption of heat and subsequent cell destruction (photothermotherapy).

and/or genome expression manipulations. Peptide-driven targeted delivery allows chemotherapeutic drugs to be delivered specifically to cancer cells, assuming that the integrity of the delivery vehicle remains intact until its final destination is reached. In some cases, the originally selected phage display clones are used for drug delivery if they can be loaded with the molecule of interest via postselection chemical modification. Cancer-specific peptides have been engineered to attach on the surface of gold nanoparticles for the purpose of photothermotherapy. The modified nanoparticles are injected in tumor locations and selectively retained there. When tumors are irradiated with lasers, the gold nanoparticles are heated and cells in their vicinity killed due to overheating. Normal cells are not harmed since the gold nanoparticles selectively attach only to cancer cells.

Targeted delivery to specific organs and tissues is another hot area of research being addressed by phage display. A library displaying

FIGURE 5.11 **In Vivo biopanning.** Phage display library is injected into a live mouse and the animal sacrificed after several hours. Organs, tissues, or tumors of interest are isolated and perfused to remove nonspecific binders. Specific binders are eluted and sequences of displayed peptides identified. Selected peptides are used for research or therapeutic applications.

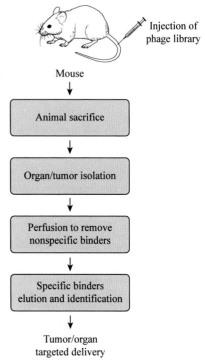

random peptides is injected in a live animal (Fig. 5.11) and after a while the animal is sacrificed, tissues and organs are isolated and perfused to remove nonspecifically absorbed phages. Next, the stably bound phages are eluted, amplified, and sequenced to identify potential organ-targeting peptides and subject them to maturation and validation. The described strategy is rather tedious and can easily be visualized as looking for a needle in a haystack; however, any selected peptide has a huge potential to revolutionize drug delivery, as well as gene therapy in the relevant organ. Peptides able to cross the brain–blood barrier are of special interest since current medicine has very limited options to deliver drugs in the brain and availability of such means is critical for breakthroughs toward management and cure of neurodegenerative diseases.

Phage display is widely used to select for peptide enzyme inhibitors and allosteric regulators, receptor agonists, and antagonists, antitoxins, etc., as well as peptides binding to specific surfaces, structures, and chemical compounds. Examples of use of phages and viruses in general in developing nanomaterials, imaging probes, and nanodevices are discussed in Chapter 6.

5.4.3 Phage Display Applications Based on Whole Phage Particles

While the number of proteins and peptides selected by phage display being used toward biomedical and biotechnological applications is rapidly growing and is being considered the main focus of the technique, more and more examples of the utility of selected phage clones are being reported. For example, phages displaying peptides stimulating cell growth and proliferation are being explored as tissue scaffolding. Tissue scaffolds are being developed to mimic the overall structure and properties of the extracellular matrix (ECM). ECM is a complex system of fibril proteins with diameters ranging between 3 and 20 nm, which provides mechanical support to cells and is able to bind soluble physiologically significant molecules essential for cell adhesion, migration, growth, and differentiation. Viruses are considered potential valuable structural elements of biomimetic tissue scaffolds since they measure on the nanometer scale, could align in orderly structures (including fibers) at high concentrations, and could be modified to immobilize various molecules. M13 and phage display are of particular interest due to the advanced understanding of the system and the options of simultaneous display of multiple molecules. In one example, tissue scaffolds were developed by simultaneously displaying an integrin-binding peptide RGD (Arg-Gly-Asp) and HPQ (His-Pro-Gln) peptide binding streptavidin. The RGD peptide was expected to anchor cells to the phage surface mimicking cell–ECM interactions, whereas the HPQ peptide was expected to serve as a velcro for growth and differentiation factors chemically conjugated to streptavidin (Fig. 5.12). Streptavidin/FGFb (basic fibroblast growth factor) and Streptavidin/NGF (nerve growth factor) were used in proof of concept experiments with neural progenitor cells (NPCs) isolated from rat hippocampus. Various display configurations were tested utilizing high density pVIII display on the major capsid protein and multivalent pIII display. The RGD and HPQ peptides were displayed individually on separate phages or together on the same phage. Mixtures of displaying and wild-type phages were also tested. The experiments demonstrated that both adhesion signals and growth signals are needed for optimal cellular growth. In the absence of the RGD peptide, cells were clumping together. The density of dividing NPCs was found to correlate with the amount of immobilized FGFb as growth was promoted to a larger extent by M13 displaying HPQ peptide on pVIII (HPQ8 phage) than on pIII (HPQ3 phage), displaying 2700 versus 3–5 copies, respectively. The dual display phage RGD8HPQ3 supported cell spreading and growth well. When NGF conjugated to streptavidin was supplied to matrices of any of the three phages described above, cell differentiation was observed. Technically, the scaffolds were assembled by

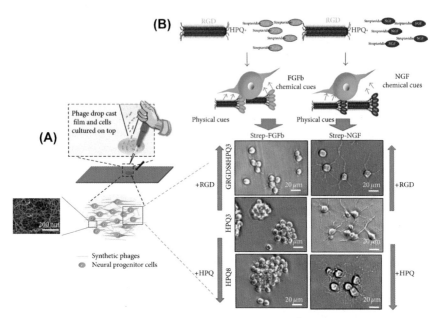

FIGURE 5.12   **Example of phage-based tissue engineering approach utilizing peptide display.** (A) Assembly and electron microscopy visualization of tissue scaffolds mimicking extracellular matrix. A droplet of high titer M13 stock was applied to a glass slide forming a film of nanofibrils as seen with scanning electron microscope (inset on the left). Neural progenitor cells were overlaid and cultured on top of nanofibrils. (B) Overview of cell proliferation and differentiation on synthetic scaffolds generated from phages displaying RGD and HPQ (streptavidin-binding) peptides, conferring cell adhesion and growth factor immobilization properties, respectively. HPQ8 phage displayed the HPQ peptide as a fusion to the major capsid protein allowing large amounts of growth factors to be attached, HPQ3 displayed the same peptide as a fusion to the minor capsid protein pIII allowing attachment of limited number of growth factor molecules and the GRGDS8HPQ3 phage displayed large numbers of RGD peptide and limited numbers of growth factor attachment sites. Immobilized FGFb (basic fibroblast growth factor) stimulated cell proliferation, whereas NGF (nerve growth factor) stimulated cell differentiation. Cell proliferation in absence of FGFb resulted in cell clumping. *Image courtesy of Yoo SY, Merzlyak A, Lee SW. Synthetic phage for tissue regeneration.* Mediators Inflamm *2014;**2014**:192790.* https://doi.org/10.1155/2014/192790, *CCBY,* https://creativecommons.org/licenses/by/3.0/.

placing a drop of phage preparation on a glass slide and overlaying the formed film with NPCs. Similar phage display strategies have been tested for other biologically active peptides relevant to cell adhesion, angiogenesis inhibition, cell proliferation inhibition, prevention, and treatment of Chlamydia, among others. It was proposed that phages displaying autoantigens to be injected in the body as a means of sponging up autoantibodies. Biotinylated bacteriophages were "decorated" with quantum dots using streptavidin/biotin conjugation and are proposed to be used as sensors and tracers. Similarly, bacteriophages attached to magnetic beads or metal surfaces utilizing displayed peptides are envisioned to give rise to sensitive detectors for pathogenic bacteria utilizing the natural affinity of phages for their hosts.

A fascinating story with many interesting twists gravitating around phage display is the development of GAIM technology (Proclara Biosciences), which in 2017 was in phase I clinical testing as a tool to combat Alzheimer's disease (AD) and potentially other neurodegenerative diseases associated with amyloid plaque formation. GAIM (General Amyloid Interaction Motif) technology utilizes GAIM protein fragment from M13 pIII capsid protein fused to a human Fc antibody backbone and is delivered intravenously. The GAIM technology is expected to clear existing Aβ plaques and tau fibrils associated with AD via the plaque depolymerization action of the two N-terminal domains of pIII. The protein fragment also binds and depolymerizes protein aggregates of alpha-synuclein (associated with Parkinson's disease) and yeast prion Sup35. Interestingly, the observation that M13 can disrupt amyloid assemblies was initially made in phage display experiments designed simply to narrow down the epitope of an antiamyloid antibody capable of disruption of amyloid plaques. Detailed studies revealed that the M13 property is due to the characteristics of its pIII protein whose natural function is to bind to the F-pili of the *E. coli* host, which serve as a primary receptor, thus mediating phage absorption. In search of tools to combat AD, M13 displaying the EFRH peptide (the epitope of antiamyloid antibody) was explored as a tool for vaccination, hoping to induce antiamyloid plaque immune response. Another filamentous phage, f88, was explored to deliver antiamyloid scFv antibody fragments to the brain evaluating the potential for live imaging of amyloid plaques in the mouse model for AD.

In conclusion, although phage display was initially developed as a ligand-mining tool, providing a straightforward link between phenotype and genotype, over time it grew into a platform with versatile practical applications in research, biomedicine, and technology and will likely continue to advance these fields in the future.

Further Reading

1. Bowley DR, Jones TM, Burton DR, Lerner RA. Libraries against libraries for combinatorial selection of replicating antigen-antibody pairs. *Proc Natl Acad Sci USA* 2009;**106**(5):1380–5.
2. Esvelt KM, Carlson JC, Liu DR. A system for the continuous directed evolution of biomolecules. *Nature* 2011;**472**(7344):499–503.
3. Fernandez-Gacio A, Uguen M, Fastrez J. Phage display as a tool for the directed evolution of enzymes. *Trends Biotechnol* 2003;**21**(9):408–14.
4. Frei JC, Lai JR. Protein and antibody engineering by phage display. *Methods Enzymol* 2016;**580**:45–87.
5. Galan A, Comor L, Horvatic A, Kules J, Guillemin N, Mrljak V, Bhide M. Library-based display technologies: where do we stand? *Mol Biosyst* 2016;**12**(8):2342–58.

6. Garufi G, Minenkova O, Passo CL, Pernice I, Felici F. Display libraries on bacteriophage lambda capsid. In: *Biotechnology annual review*, vol. 11. Elsevier; 2005. p. 153–90.

7. Konning D, Zielonka S, Grzeschik J, Empting M, Valldorf B, Krah S, Schroter C, Sellmann C, Hock B, Kolmar H. Camelid and shark single domain antibodies: structural features and therapeutic potential. *Curr Opin Struct Biol* 2016;**45**:10–6.

8. Krumpe LR, Mori T. T7 lytic phage-displayed peptide libraries: construction and diversity characterization. *Methods Mol Biol* 2014;**1088**:51–66.

9. Makela AR, Oker-Blom C. Baculovirus display: a multifunctional technology for gene delivery and eukaryotic library development. *Adv Virus Res* 2006;**68**:91–112.

10. Van de Broek B, Devoogdt N, D'Hollander A, Gijs HL, Jans K, Lagae L, Muyldermans S, Maes G, Borghs G. Specific cell targeting with nanobody conjugated branched gold nanoparticles for photothermal therapy. *ACS Nano* 2011;**5**(6):4319–28.

6

Viruses as Nanoparticles

6.1 NANOTECHNOLOGY OVERVIEW

Nanotechnology is an interdisciplinary field at the crossroads of chemistry, physics, molecular biology, engineering, and material sciences focused on creating high-performance products relevant to multiple aspects of life, ranging from biomedical nanodevices for drug delivery to superperformance tennis rackets, water-repelling fabrics, and invisible sunscreen. The field is broadly defined as "the manipulation of matter with at least one dimension sized from 1 to 100 nanometers" (US National Nanotechnology Initiative), thus covering essentially any practical applications of any virus falling in that size range. Nanotechnologies should not be confused with microtechnologies, which are focusing on designing miniature versions of existing products, although some applications can easily fit both definitions. For example, miniature batteries based on metal-covered virus capsids smaller than 100 nm technically can be considered a nanotechnology product due to size, as well as a microtechnology product since they are a miniature version of a product already in use in our macroworld.

On the scale of natural sciences, nanotechnology is a young field, which emerged in the second half of the 20th century and has been developing with superfast pace ever since. The discovery of the fullerenes, carbon-based structures with a 3-D hollow shape, and the invention of high-resolution microscopes (atomic force microscope and the scanning tunnel microscope) are often considered the cornerstones of the field. Looking back at the history of human civilization with our current understanding of science, one can find empirical examples of practical applications that are technically nanotechnologies, such as metal-based stained glass, which in fact employs metal nanoparticles, or the so-called Damascus saber blades owning their high quality to the properties of what would be described today as nanotubes and nanowires. Currently, the web portal of the US National Nanotechnology Initiative maintains an interactive timeline of discoveries and products related to nanotechnologies. If one is visiting a science museum, the idea of nanoparticles and nanotechnology is most likely introduced using the example of colloidal gold changing color from gold to red. As a young and fast-growing field, nanotechnology is also at the crossroads of policymaking and establishing safe usage standards (including both health and environmental ones), a process further complicated by the diversity of products and approaches to manipulate matter on the nanoscale. The future of the field is often discussed in the context of two major trajectories: new materials and complex nanodevices, many of which are geared toward biomedical applications. Vaccines, drug and gene delivery, and biomedical imaging are among the areas of virus-related research that are expected to be most influenced by nanotechnology.

6.2 VIRAL CAPSIDS AS A COMMODITY FOR NANOTECHNOLOGY

From the perspective of virology, one can argue that the development of conventional whole virus vaccines is among the first "newborns" of nanotechnology. For example, the inactivated polio vaccine is made by manipulating viral particles with icosahedral symmetry roughly 30 nm in diameter, thus falling in the lower half of the size range relevant to nanotechnology. As great as this historic example is, it addresses a single application of viral capsids as an immunity-inducing entity and hints at only a couple of the reasons why viral capsids are extremely helpful for developing nanotechnologies: they are easy to produce and straightforward to destroy. Viral capsids are much more versatile than that, and their utility has much to do with their biological nature, the many years of coevolution with their hosts, and extensive research on their properties and characteristics translating into a broad array of tools and options for modification. The current discussion will be limited to four major aspects of viral capsids' value to nanotechnology: (1) accessibility, (2) biocompatibility, (3) ability to self-assemble, and (4) potential for modification.

6.2.1 Viral Capsid Accessibility

Viruses are a very diverse and very abundant biological entity found essentially everywhere in nature. Most likely no one will ever know exactly how many different virus species or how many total viruses are on the planet, simply because those numbers are dynamic; however, it is easy to be relatively precise describing the number of viruses in nature as tremendous. It has been estimated that there are 10^{31} viruses on Earth, mostly bacteriophages. Furthermore, viruses have been isolated from very diverse environments in terms of pH, temperatures, salt concentration, pressure, etc. and are known to infect organisms from all kingdoms.

Therefore, theoretically there is a huge pool of viral capsid types with different properties available to nanoscience. Naturally, current efforts are focused on the best-studied viruses and, understandably, the most useful ones are bacteriophages and plant viruses, which do not pose health risks to humans or economically significant animals (Fig. 6.1). Accessibility is also a function of our ability to produce large amounts of high-quality isomorphic particles in a reasonable time and at a reasonable cost. Virus nanoparticles (VNPs) are essentially viral capsids, which are produced "naturally" by allowing viruses to propagate in their hosts. VNPs have limited applications, unless they are modified/engineered to acquire special properties. Most frequently, VNPs are genetically engineered to express proteins/peptides of interest, which confer functionality or allow other molecules to be attached and do so. Virus-like particles (VLPs) are of better use since they lack the original viral genome and thus cannot be

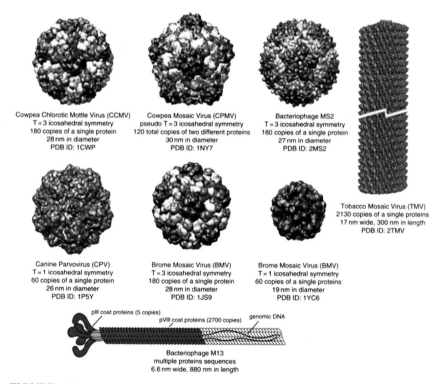

FIGURE 6.1 **Examples of viral capsids commonly considered for nanotechnology applications.** Viral capsids are a valuable resource for nanotechnology purposes due to their small size, regular structure, and potential for modifications, accessibility, and biocompatibility. *Reprinted from Stephanopoulos N, Francis MB. Making new materials from viral capsids. In: Matyjaszewski K, Möller M, editors.* Polymer Science: A Comprehensive Reference. *Amsterdam: Elsevier; 2012. p. 247–66, with permission from Elsevier.*

FIGURE 6.2   **pH-dependent assembly of cowpea chlorotic mottle virus (CCMV) cap-sid.** CCMV disassembles to viral RNA and CP dimers at pH 7.5 in the presence of $CaCl_2$. Following the removal of the viral genome, the empty capsid can be reassembled by chang-ing the pH of the solution to 5. *Reprinted from Ma Y, Nolte RJM, Cornelissen JJLM. Virus-based nanocarriers for drug delivery.* Adv Drug Deliv Rev 2012;**64**(9):811–25, *with permission from Elsevier.*

infectious. VNPs and VLPs of bacterial, animal, and plant viruses have been successfully produced in laboratory and commercial settings using various in vivo and in vitro approaches. VNPs are produced in their natu-ral host, whereas VLPs are mainly produced in heterologous expression systems, most frequently *E. coli*, yeast, or plants. Insect or mammalian tis-sue culture expression systems are also viable options. For many viruses, empty capsids are by-products of the natural assembly process and can be isolated from infected cells in parallel with VNPs. If feasible, VLPs can also be produced from VNPs by removing their nucleic acids. The lat-ter could be accomplished by pH-driven swelling of VNPs followed by alkaline hydrolysis, which essentially release and chemically destroy the viral nucleic acids. Alternatively, VNPs could be forced to disassemble to individual capsid proteins (CPs), the nucleic acids could be extracted, and the CPs reassembled into the VLP. For example, CCMV (cowpea chlorotic mottle virus infecting black-eyed pea/bean plants) capsid can be disas-sembled to CP dimers and RNA at pH 7.5 in the presence of calcium salt and assembled as an empty VLP at pH 5 (Fig. 6.2). Both approaches of nucleic acid removal require coat proteins to be able to "survive" the treat-ment and readily associate with each other to reform a robust capsid.

6.2.2 Viral Capsid Biocompatibility

Viral capsids are naturally biocompatible with living organisms and superior to any synthetic nanoparticles. They are stable enough to be used as a delivery vehicle and can be cleared and/or degraded by the body with considerable rates, thus making them a convenient tool with lim-ited chances for side effects. Research shows that virus-based nanopar-ticles have half-lives on the scale of minutes to hours, in sharp contrast to unmodified synthetic nanoparticles, some of which were found to be selectively retained in liver or kidney of experimental animals. For both synthetic and virus-based nanoparticles, persistence in the body has been correlated to charge. For example, negatively charged VLPs derived from cowpea mosaic virus (CPMV) and CCMV have circulation half-life less than 15 min. In contrast, the positively charged capsid of bacteriophage Qβ has a half-life longer than 3 h. Theoretically, the immunogenicity of viral capsids may present a hurdle for extensive repetitive applications of the same type of capsid, as immune response can diminish their effectiveness due to speedy neutralization. Practically, very little is known about the immunogenicity of the most common VLPs and VNPs considered for applications in humans. The combination of short half-life and active immune response might significantly narrow the time window of persistence/circulation in the body. It is possible that in the near future, VLPs used as carriers to deliver heterologous antigens for the purpose of vaccination may become a regular part of our vaccination records to ensure vaccine effectiveness. Furthermore, a need may arise for global agreements to regulate which VLPs are to be used in support of a certain vaccination effort, so that immune response against a VLP already used for vaccination does not diminish the chances of a subsequent vaccination event. Immune response to viruses used for biomedical applications is an important issue in the context of phage display applications, bacteriophage therapy, gene therapy, and biomaterials. Very likely this issue will attract significant research efforts as more virus-based applications show promise in the lab and enter the pipelines of clinical testing.

6.2.3 Ability to Self-Assemble

Self-assembly is a hallmark of viral capsids, shaped by host/virus interactions and evolution. At the current state of nanoscience, viral capsids are much more cost-effective to produce in large quantities and with higher quality than any synthetic nanoparticle, thus making self-assembly a rather critical property. Capsid assembly of icosahedral and complex viruses is driven by protein–protein interactions and takes place naturally if coat protein(s) are expressed in sufficiently high concentrations and correct ratios in an environment promoting self-association. Assembly of helical/filamentous viruses is driven by protein–protein and protein–nucleic acid interactions. Remarkably, some capsids and even infectious viruses can be reconstituted in a test tube from purified components. Several examples of VNPs and VLPs assembly with a different level of complexity are discussed below in an attempt to present the state of the field and the type of expression/assembly challenges being solved.

A VLP-based biomedical application, which has been well accepted for years worldwide is the anti-HPV (human papillomavirus) vaccine recommended as a preventive measure against cervical cancer and papilloma infections in general. The major CP of HPV, L1, can self-assemble in penton capsomers (Fig. 6.3), 72 of which can form a stable VLP eliciting a robust immune response. The described VLP is widely used as an

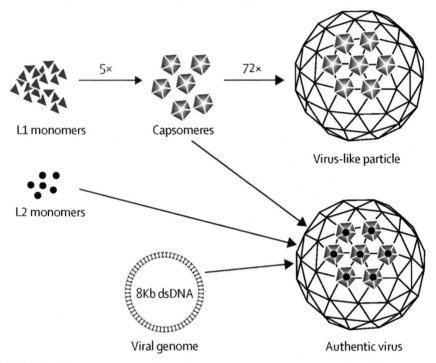

FIGURE 6.3 **Structure of human papilloma virus (HPV) and human papilloma virus–viruslike particle (VLP).** HPVs are naked dsDNA viruses with circular genome and icosahedral capsids composed of major (L1) and minor (L2) capsid proteins (bottom of figure). They are widely spread in human population causing asymptomatic infections; warts and some strains/types can cause cancerous growth. Anti-HPV vaccines based on VLP composed of the major capsid protein, L1 (top of the figure), are currently in use marketed as vaccines against cervical cancer. *Reprinted from The Lancet Oncology, Schiller JT, Müller M. Next generation prophylactic human papillomavirus vaccines. Lancet Oncol 2015;16(5):e217–25. Copyright (2015), with permission from Elsevier.*

injectable vaccine preparation conferring anti-HPV immunity. In fact, the HPV VLP could be characterized as a subcapsid particle, since it is lacking not only the viral genome but also the minor capsid protein, L2. L2 is critical for viral entry in the cell during infection as well as for protein–protein interactions with cellular proteins involved in intracellular transport (beta actin and syntaxin). The heterologous expression of functional L1 protein in *E. coli* and yeast makes the production of L1 VLP cost-effective and alleviates safety concerns. The stability and the immunogenicity of the L1 VLP in the absence of the minor CP L2 allow for straightforward expression and assembly process. Prior knowledge regarding virus biology and the clear understanding of the functions of each CP secured a great level of confidence for the safety of the vaccine before clinical testing took place. Despite the "apparent simplicity" of the VLP-based assembly, the anti-HPV vaccine is not without its challenges. Unfortunately, many strains/types of HPVs circulate in nature and pose a threat to human health. Multiple anti-HPV vaccines exist on the market, offering protection against a variable number of HPV strains.

Tobacco mosaic virus (TMV) is an example of a helical virus that can self-assemble in a test tube. Although that has been known for more than half a century, the advances of nanotechnology renewed the interest in the process as an accessible platform for the development of nanomaterials. The TMV virion is a hollow tube 300 nm long and 18 nm in diameter with an internal channel measuring 4 nm in diameter. TMV assembly is a complex process guided by its RNA (Fig. 6.4). A special sequence designated as origin of assembly or "OA" sequence, located toward the 3′ end of the molecule, is essential for the initiation of the process. The OA sequence interacts with a self-assembled disk-like structure of CP. The interaction results in a "lock washer" structure that tethers the RNA molecule to the disk and creates means for threading the rest of the RNA, commonly described as an RNA traveling loop. Capsid polymerization takes place bidirectionally. The capsid grows toward the 5′ end of RNA by adding protein disks and toward the 3′ end of RNA by adding small clusters of CP designated as "A-protein." Preassembled protein units are continuously added until the entire RNA length is covered. Engineered TMV genomes with various lengths are readily packed, thus providing an opportunity for generation of nanotubes with different sizes.

A huge advantage of the availability of in vitro assembly reactions is the opportunity they create for fast-paced experimentations to understand mechanisms of assembly and to apply that knowledge toward exploration of controlled assembly/disassembly reactions. In vitro assembly of VNPs and VLPs creates opportunities for detailed studies of the self-assembly processes, exploration of approaches for capsid modifications, and, hopefully in the near future, mechanisms for capsid assembly/disassembly on demand; however, so far, they have been possible for very few viral systems employing a single CP. In viruses with complex capsids or sophisticated assembly processes, the production of VNPs or VLPs outside the context of infection of the natural host is a task to be accomplished in the future.

Engineering of replication-defective helper viruses or introducing essential viral functions into the host genome are viable strategies to produce VNPs with desired properties. Helper M13 viruses are routinely used to package M13 phagemid vectors in M13 capsids for the purpose of phage display. M13 has a complex filamentous capsid built of several thousand copies of its major CP and decorated on the tips by minor CPs (Fig. 6.1 and Chapter 5). The length of the filament is proportional to the length of the packaged DNA, thus allowing engineering of

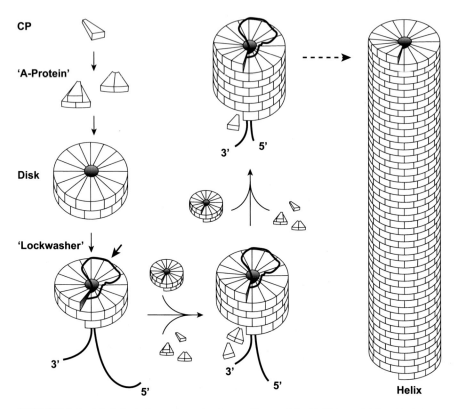

FIGURE 6.4   **Model of tobacco mosaic virus self-assembly.** TMV is a ssRNA naked virus composed of one RNA molecule depicted in black (6395 nucleotides long) and 2130 molecules of a single coat protein. In a test tube the coat protein forms two stable multimeric structures: "A protein" and disks. The assembly starts with disk attachment to a specific sequence called origin of assembly (OA) resulting in a "lock washer" and the formation of an RNA "traveling loop" (pointed with an arrow). Capsid assembly proceeds bidirectionally as "A protein" assemblies are added toward the 3′ end of RNA and disk assemblies are added toward the 5′end of the RNA. *Image courtesy of Koch C, Eber FJ, Azucena C, Förste A, Walheim S, Schimmel T, Bittner AM, Jeske H, Gliemann H, Eiben S, Geiger FC, Wege C. Novel roles for well-known players: from tobacco mosaic virus pests to enzymatically active assemblies.* Beilstein J Nanotechnol 2016;7:613–29. https://doi.org/10.3762/bjnano.7.54; CCBY2.0 – http://creativecommons.org/licenses/by/2.0.

different size nanoparticles. The phagemid vectors contain the DNA to be packaged, whereas the genetic information for the CPs is introduced in the cell manufacturing nanoparticles *in trans* from the genome of a replication-defective helper virus. A similar strategy is used to package recombinant baculoviruses for protein expression in insect cells (Chapter 4); however, in this case, the genetic information for virion packaging is contained on a bacmid previously inserted in the packaging cell. Not surprisingly, in addition to protein expression, recombinant baculoviruses are being utilized for baculovirus surface display and as vectors for gene therapy.

Assembly of multicomponent VLPs can be a challenging process, as expressing multiple proteins together in the correct ratio and in high concentration pushes the limits of the available expression systems. Good understanding of the basic biology of the virus of interest can provide useful tricks for facilitating VLP production. For example, tobacco coexpression of the coat protein, precursor protein, and viral protease of CPMV results in assembly of a stable VLP (Fig. 6.5). CPMV capsid is built from equimolar amounts of large and small coat protein synthesized as one long polypeptide, which is cleaved by a virus-encoded protease. In this case, simply taking advantage of an evolutionarily established mechanism to control the availability of both CPs in a one-to-one ratio delivers the desired result. Recently, a VLP based on bluetongue virus, BTV, was also expressed and assembled in tobacco (Fig. 6.5). BTV causes bluetongue disease in ruminants and has a major economic impact, thus production of BTV VLPs is of huge interest for the purpose of livestock vaccination. BTV's capsid is built from four different coat proteins arranged in three protein layers surrounding the dsRNA segmented genome of the virus packaged together with several nonstructural proteins. The sophisticated capsid structure is a hallmark of reoviruses and its complexity can easily be rationalized by the chemical composition of the viral genome. dsRNA is an "unwelcome" molecule in the eukaryotic cell. Once detected, dsRNA is quickly destroyed; thus to propagate any dsRNA, the virus has to evolve mechanisms to protect its genome from detection and destruction.

The outer capsid of the BTV is composed of 60 VP2 trimers and 120 VP5 trimers and is essential for BTV entry in the cell. The removal of the outer capsid releases a core particle, which is transcriptionally active and supports the initial steps of BTV replication in a cell-independent manner, i.e., one can think about it as an autonomous enzyme-driven factory or reactor. The outer layer of the core particle is composed of 260 trimers of the VP7 CP, whereas the inner layer is composed of 60 dimers of the VP3 protein. Core-like particles have been assembled in insect cells infected with recombinant viruses coding for VP3 and VP7 proteins alone or in combination with baculoviruses coding for viral enzymes essential for replication. Historically, the outer capsid studies have been performed on virions isolated from infected mammalian cells. Surprisingly, despite the complexity of the BTV capsid, VLPs were successfully assembled in tobacco plants when all four CPs were expressed together using the CPMV-HT (cowpea mosaic virus–hyper translation) system (Fig. 6.5 and Chapter 4). The 5' UTR of the VP3 gene was mutated to modulate its expression relative to the other three CPs, thus reducing the amount of capsid intermediates and ensuring a high yield of structurally sound BTV VLPs.

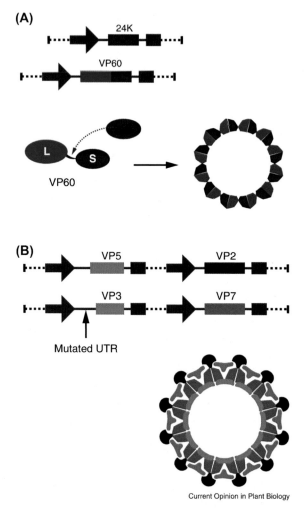

Current Opinion in Plant Biology

FIGURE 6.5 **Strategies for production of multicomponent viruslike particles (VLPs).** (A) CPMV VLPs are produced by coexpression of a coat protein precursor, a polyprotein composed of the large and small CP, and protease cleaving the precursor into individual polypeptide chains. The approach ensures that both capsid proteins are expressed in equimolar amounts. (B) BTV VLPs are produced by coexpression of four capsid proteins. The 5′ UTR of one of the proteins is mutated to secure lower expression to match the natural stoichiometry of the capsid proteins. *Reprinted from Sainsbury F, Lomonossoff GP. Transient expressions of synthetic biology in plants.* Curr Opin Plant Biol *2014;19:1–7, with permission from Elsevier.*

6.2.4 Potential for Modifications to Meet Diverse Functionalities

As protein-based structures, capsids can be altered and their applications fitted to particular purposes using the arsenal of techniques already

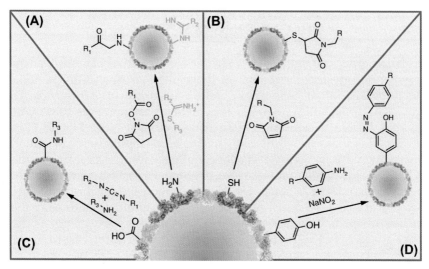

FIGURE 6.6 **Widely used approaches for chemical modification of virus nanoparticles (VNPs) and viruslike particles (VLPs).** Commonly modified side chains on both internal and external capsid surfaces include: (A) amino groups (lysine, *N*-terminus) modified by NHS-ester conjugation (left, black) or thioimidate conjugation (green, right). (B) sulfhydryl groups (cysteines) modified by maleimides or disulfide bonds. (C) carboxylic groups (glutamic acid and aspartic acid) modified by primary amines. (D) phenol groups (tyrosine oxidation to diazonium permits for ortho attachment to the phenol group of the side chain). *Reprinted from Smith MT, Hawes AK, Bundy BC. Reengineering viruses and virus-like particles through chemical functionalization strategies.* Curr Opin Biotechnol 2013;24(4):620–6, *with permission from Elsevier.*

available to modify proteins. These techniques include: (1) genetic modification; for example, insertion, removal, or altering of individual amino acids, protein segments, or entire protein domains; (2) covalent or noncovalent chemical modifications on the outer or inner capsid surface to create possibilities for stable attachment of a molecule of interest; and (3) modifications to alter charge and surface immunogenicity.

Covalent modifications of capsids are based on well-established methods for cross-linking, fluorescent or paramagnetic labeling of proteins, and most frequently using chemistries utilizing amino-, carboxyl-, and sulfhydryl-containing side chains (Fig. 6.6). Noncovalent modifications are based on electrostatic, protein/protein, protein/nucleic acid, or protein/metal interaction. The diversity of available linkers, fluorescent dyes, etc. makes virtually every capsid somewhat modifiable, assuming enough experimentation is done to ensure preservation of functionality and acquisition of the desired properties. Limited options for modification are available to reduce capsid immunogenicity. PEG (polyethylene glycol) conjugation is most frequently used for that purpose.

6.3 NANOMOTORS: ELEGANT NUCLEIC ACID PACKAGING MACHINES

Nanomotors are molecular machines that utilize chemical energy to generate physical movement of molecules. Viruses utilize nanomotors to package their genomes in preformed capsids or procapsids. The process and machinery required for genome packaging are best understood in dsDNA bacteriophages, such as Lambda, P1, P2, T4, and T7, all infecting *E. coli*, as well as *Bacillus subtilis* bacteriophages Phi29. Among eukaryotic viruses, genome cleavage and packaging is best understood in herpesviruses, also with dsDNA genome. Comparative analysis has identified a number of conserved functions required for threading DNA into procapsids. One can think of procapsids as empty capsids since they are lacking nucleic acid; however, they are not physically empty as their inner space is full of scaffold proteins (SPs) critical for capsid assembly. Procapsids are converted into mature capsids once the genome is packaged, the scaffold removed, and in many cases conformational changes take place. The functions of the following proteins are required for genome packaging: CPs, SPs, and the portal vertex packaging complex. The portal vertex packaging complex is a ring-like structure that contains a nanomotor able to bind and thread DNA through the channel of the ring. Portal vertex complexes utilize protein subunits (most viruses) or protein and packaging RNA (pRNA) in the case of the phi29 and phi29-like bacteriophages.

Two principal mechanisms of nanomotor genome packaging have been described so far: (1) packaging driven by terminal proteins and (2) packaging driven by packaging sequences/terminases (Fig. 6.7). Packaging driven by terminal proteins is best understood in bacteriophage *phi29*, which has a single molecule of terminal protein (gp3) covalently attached to the 5′ end of each DNA strand in its linear dsDNA genome. The terminal protein is an inseparable part of the phage genome and is also used for replication priming providing a serine hydroxyl group to the *phi29* DNA polymerase to start synthesizing DNA. After replication is completed, the terminal protein on the left side of the genome forms a small lariat structure, which is recognized by the portal complex and the left end of the genome is inserted into the procapsid. The nanomotor of the packaging complex eventually pulls in the entire genome, completing the packaging process. Packaging driven by DNA viral sequences/terminases has been described in dsDNA viruses that replicate by the rolling circle mechanism and produce long head-to-tail concatemers. The packaging complex (specifically the small terminase subunit) selectively recognizes and binds to the packaging sequence, then threads the genomic DNA into the procapsid until the next packaging sequence is encountered. DNA is cut at genome length and the portal

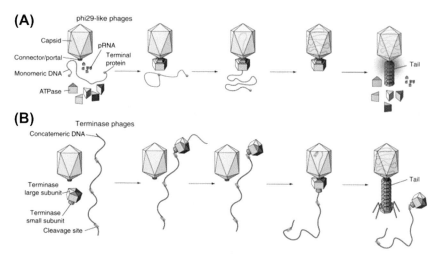

FIGURE 6.7   **Genome packaging mechanisms of dsDNA bacteriophages.** (A) Terminal protein–driven genome packaging. Bacteriophage phi29 genome, "marked" with the terminal proteins (*orange balls*), forms a lariat and is bound by the portal packaging complex (ATPase [blue], connector [green] and pRNA [magenta]) attached to the procapsid (yellow). Once packaging is completed, the ATPase and pRNA dissociate and the tail complex (gray) attaches. (B) Packaging sequences/terminase-driven genome packaging. Genome packing sequences (*orange boxes*) guide the terminase complex (large subunit [blue] and small subunit [red]) to package and cleave full-length genome. Once packaging is completed, the ATPase and pRNA dissociate and the tail complex (gray) attaches. *Reprinted from Hetherington CL, Moffitt JR, Jardine PJ, Bustamante C. Viral DNA packaging motors. In: Egelman, Edward H, editors.* Comprehensive Biophysics. *Amsterdam: Elsevier; 2012. p 420–46, with permission from Elsevier.*

complex dissociates with the reminder of the concatemer, ready to attach to a new procapsid and repeat the process. In tailed phages, DNA packaging by both mechanisms is followed by tail assembly, thus resulting in a mature particle, whereas in viruses without tails, the portal vertex is plugged to prevent genome exit.

Packaging/portal complexes are ring structures, which contain a portal/connector and an ATP-driven nanomotor. The connector is comprised of 12 subunits building a funnel-like structure. The wide end of the funnel is "inserted" into the procapsid, whereas the narrow end serves as an interface for interaction with the nanomoter subunits. The internal channel of the connector guides the DNA being packaged into the procapsid. The ATP-driven motor is also oligomeric, although the number of subunits varies among viruses. In terminase packaging phages, both ATP-driven DNA movement and the nucleolytic cleavage at the end of the genome packaging are carried out by the large terminase subunit. The bacteriophage *phi29* packaging complex employs a unique pRNA (Fig. 6.8). pRNA is encoded by the left end of the genome and produced

(A)   **(B)**   **(C)**   **(D)**

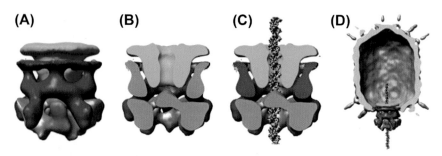

FIGURE 6.8   **Structure of phi29 molecular nanomotor.** The phi29 molecular nanomotor consists of connector (green), pRNA (magenta), and ATPase (blue): (A) side view, (B) cross-section with the front half removed, (C) cross-section of a nanomotor treading DNA through the central channel, (D) nanomotor in the context of the entire capsid. *Reprinted from Morais MC, Koti JS, Bowman VD, Reyes-Aldrete E, Anderson DL, Rossmann MG. Defining molecular and domain boundaries in the bacteriophage ϕ29 DNA packaging motor. Structure 2008;16(8):1267–74, with permission from Elsevier.*

as a 174 bp transcript with a complex secondary structure. The pRNA ring is assembled on the procapsid as the last step of its morphogenesis. pRNA is required for the nanomotor assembly and function. It has been found to stimulate ATP hydrolysis and to assist with packaging initiation. Once packaging is completed, the pRNA is released together with the nanomotor prior to the addition of the tail components. The precise modes of action of packaging nanomotors of both types are not completely understood. Simplistically, one can describe DNA threading as an ATP hydrolysis–driven process associated with protein conformational changes resulting in stepwise unidirectional movement of DNA in a sequence-independent manner. Single molecule techniques are expected to contribute to a detailed understanding of the process and facilitate practical applications of packaging complexes. In addition to the motor-related applications, the *phi29* packaging complex is considered as a possible module of nanosensors, based on the occupancy of the DNA translocation channel (Fig. 6.9). When the connector channel is inserted into lipid bilayer and electrodes applied to both sides of the membrane, its conductivity can be monitored and correlated to the presence or absence of molecules (DNA, ions, etc.) in the channel cavity. When epithelial cell adhesion molecule (EpCAM) peptide was fused to the C-terminal end of the connector subunit, it was able to detect the presence of bound anti-EpCAM antibody in the channel of the connector, thus demonstrating potential utility as a probe in biomedical diagnostics. The system was found to be sensitive and specific as demonstrated by the detection of nanomolar concentrations of specific antibodies and discrimination against nonspecific antibodies in serum samples.

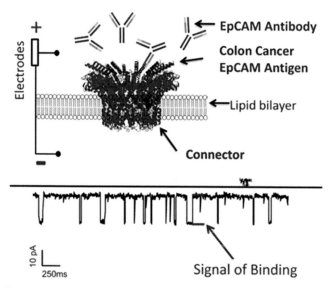

FIGURE 6.9   Application of phi29 packaging complex in nanosensors. The phi29 connector displaying short epithelial cell adhesion molecule (EpCAM) peptide is embedded into a lipid bilayer with electrodes attached on both sides. Anti-EpCAM antibodies are flown over the surface of the connector and their attachment to the antigen peptide monitored by changes in conductivity. EpCAM is a diagnostic antigen associated with colon cancer. *Reprinted with permission from Wang S, Haque F, Rychahou PG, Evers BM, Guo P. Engineered nanopore of Phi29 DNA-packaging motor for real-time detection of single colon cancer specific antibody in serum. ACS Nano 2013;7(11):9814–22. Copyright (2013), American Chemical Society.*

6.4 KEY APPLICATIONS OF VIRUS-BASED NANOTECHNOLOGIES

The versatile properties of viruses make them an attractive choice for development of new nanoscale technologies pertinent to various fields. Biomedical applications include (but are not limited to) fields related to immunotherapy, gene delivery, drug delivery, tissue engineering, vaccine development, bioimaging. Biotechnological applications involve phage display for identification, selection, and analysis of a broad spectrum of molecules of interest, development of sensor systems based on the repetitive nature of capsid subunits, energy-related applications such as battery electrodes, etc. Some of these applications are discussed in detail in other chapters: phage display (Chapter 5), vaccines (Chapter 8), therapeutic approaches (Chapter 9). The rest of this chapter will focus on a general overview of applications and a limited number of proof of principle examples, rather than extensive inventory of ideas and development efforts. On the big picture level, most VLPs and VNPs serve their purpose by functioning as nanodelivery vehicles, nanobioreactors, imaging probes, or building blocks of new materials and/or nanodevices.

6.4.1 Viruslike Particles as Nanodelivery Vehicles

From the point of view of nanotechnology, VLPs can be considered sophisticated vehicles offering cargo room and display surface to be used individually or in combination to deliver a load of molecules to a specific destination. Depending on the end goal of the application, different VLP properties can prove to be critical for successful outcome. Size and shape of VLPs are directly connected to their mobility within living matter. For example, small icosahedral VLPs diffuse best in tissues but have limited cargo capacity. On the other hand, large filamentous VLPs side better with the walls of blood vessels and, correspondingly, have larger cargo capacity. It is generally thought that the physical dimensions of capsids and their stability in nature are product of evolutionary selection driven by virus infectivity and specifics of viral whereabouts; however, some capsids can be assembled in alternative sizes naturally or after modifications. M13 capsid, for example, can vary significantly in length depending on the size of the packaged DNA genome. The compactness of icosahedral capsids is at least in part determined by the protein–protein interactions along the edges of adjacent capsomers or by the availability of the assembly scaffold and its size. Various degrees of modification of the N- and the C-termini of the CP of CCMV result in assembly of various sizes and architectures of capsids: 18 nm diameter capsid built from 60 subunits, 24 and 28 nm capsid built of 120 and 180 subunits, respectively. The CCMV capsid is sensitive to changes in pH and can reversibly transition between natural and swollen conformations. The latter is characterized with increased size and opening of "pore-like structures" with diameter of 2 nm and hypothesized to mimic an in vivo uncoating intermediate. Under specific pH, salt concentration, and ionic strength, CCMV can completely disassemble or form alternative structures such as sheets and tubes. The ability to form alternative structures is not unique for CCMV. Several other plant viruses, including TMV, can transition between alternative structures. When exposed to high temperature, the TMV CP can form spheres with different diameters in a concentration-dependent manner. The TMV virion accomplishes the transition when heated to 94°C, whereas RNA-less CP assemblies transition at 65°C. Alternative structures allow for efficient development of multiple applications with the same resources (production and purification system, toxicity testing, etc.), as well as development of switches to be incorporated as functional elements in nanodevices.

The charge and the chemical nature of the inner and outer VLP surfaces are directly connected to the type of cargo that can be transported and/or displayed and the propensity of VLPs to stick nonspecifically to surfaces. For some applications, it is important to minimize the immunogenicity of VLPs or their surface charges to prevent attachment to nonspecific surfaces. In such cases, polymers (PEG) or proteins (serum albumin) can be attached to VLP surfaces, effectively masking epitopes or compensating charges.

Cargo loading and retention can be achieved passively without alteration in the chemical structure of the cargo and the nanodelivery vehicle or actively by chemical modification or genetic fusion. For example, the photoreactive molecule phthalocyanine (Pc) can be packaged passively into CCMV VLPs by simply incubating CPs and water-soluble zinc phthalocyanine (ZnPc) in a test tube. Depending on the pH of the assembly environment, ZnPc is loaded in capsids with two different sizes carrying distinct amounts of the molecule in alternative configurations driven by differences in density of packing (Fig. 6.10). The utility of CCMV VLPs loaded with ZnPc as an agent for photodynamic therapy was demonstrated in tissue culture. The exposure to ZnPc, followed by irradiation with a mercury lamp (620–660 nm wavelength), resulted in increased cell death providing proof of principle evidence for the potential of the approach. Presumably, ZnPc was released by disintegration of the VLPs, entered the

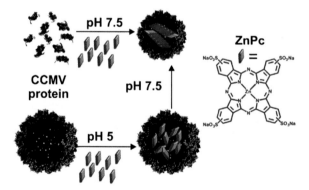

FIGURE 6.10 Passive packaging of phathalocyanine into cowpea chlorotic mottle virus (CCMV) viruslike particles (VLPs). Zinc phthalocyanine (ZnPc) is a water-soluble nontoxic porphyrin compound that is considered a promising anticancer drug for the purposes of photodynamic therapy. When irradiated with mercury lamp (620 nm), ZnPc triggers the release of reactive oxygen species, which activates cellular stress responses directing the cell to apoptosis. If CCMV capsid proteins and ZnPc are mixed in solution, the compound is packaged passively in the cavity of the self-assembling VLP. Depending on the conditions, two different sized VLPs are packaged: a larger one encapsulating ZnPc dimers and a smaller one encapsulating ZnPc stacks supposedly forming as a result of high local concentrations. ZnPc delivered to cultured cells with VLPs successfully triggers apoptosis after mercury lamp irradiation. *Reprinted with permission from Brasch M, de la Escosura A, Ma Y, Uetrecht C, Heck AJ, Torres T, Cornelissen JJ. Encapsulation of phthalocyanine supramolecular stacks into viruslike particles.* J Am Chem Soc 2011;**133**(18):6878–81. Copyright (2011), American Chemical Society.

cells, and killed them after it was photoactivated. The medical feasibility of the approach will depend on the stability of the VLPs, i.e., their ability to stay intact until their final destination, as well as their capacity to specifically target cancer cells. Passive cargo encapsulation has been shown to work well mainly for small molecules and it is presumed that it could be extended to most medically relevant small molecule drugs.

Nucleic acids present a valuable cargo for vaccine delivery and gene therapy, as well as in the context of experiments associated with gene replacement, gene knockouts, Clustered Regularly Interspaced Short Palindromic Repeats (CRISPR) genome editing, etc. They are packed electrostatically in capsids with a positively charged interior or using conventional packing mechanisms employing packing DNA/RNA sequences and the associated protein machinery (portal vertex, nanomotors, etc.). Peptide and protein cargo can be loaded passively, electrostatically, by fusion to coat/SPs building the VLP/VNP, and by chemically attaching them to coat/SP or the viral genome (specific examples are provided below). Lastly, nanoparticles, such as quantum dots or nanoparticle-based contrast imaging agents, are loaded passively or through coordination/covalent attachment to coat proteins.

Regardless of the chemical nature of the cargo and the approach to loading and retention, one of the biggest challenges for the safe and efficient use of VLPs and VNPs as nanodelivery vehicles is proper targeting to the correct destination in the body. Nonmodified nanoparticles interact with living organisms based on their natural characteristics. Data about body distribution of VLPs and VNPs are acquired in toxicity studies, which are currently starting to accumulate as more applications are moving in the direction of in vivo testing and potentially clinical trials. CCMV and CPMV, for example, show a broad distribution in multiple tissues and organs, primarily accumulating into thyroid gland and liver, respectively. Bacteriophages MS2 and Qβ were found to accumulate in the liver and the spleen. The latter is to be expected since liver and spleen function as detoxification organs. It is important to note that broad distribution and accumulation are not predictive for toxicity. In fact, CCMV and CPMV do not pose toxicity risks, consistent with their plant origin and half-life around 15 min. Nanodelivery VLPs/VNPs with broad distribution are excellent tools for systemic delivery of cargo throughout the entire body. Interestingly, CPMV exhibits a natural affinity for vimentin, an intermediate filament protein found in the cytoplasm of mesenchymal cells, as well as on the surface of endothelial cells lining up blood and lymphatic vessels. Furthermore, vimentin is overexpressed in tumors, making their vasculature easily targeted by unmodified CPMV and the cargo carried by it. VLPs or VNPs can be modified to reach destinations of interest utilizing molecules/mechanisms unique for the organ/tissue or cell type. For example, many cancer cells express a receptor for folic acid or transferrin, an iron storage protein. VLPs displaying folic acid or transferrin on their surface efficiently target the cells with cognate receptors and could potentially deliver encapsulated chemotherapeutic drugs or imaging agents straight to the target. Phage display biopanning in live animals or against surface molecules of cells in tissue culture is used to identify such key targeting molecules (Chapter 5). Approaches allowing nanodelivery vehicles to go across the blood–brain barrier are of special interest in hopes of delivering chemotherapeutic agents to inoperable brain tumors or medications relieving neuropathic pain. HIV-Tat peptide has been shown to direct molecules across the brain–blood barrier by transcytosis. Transcytosis is a specialized process, which moves molecules across cellular bodies using vesicles (Fig. 6.11). One can visualize transcytosis as back-to-back endocytosis and exocytosis. Cargo displaying the HIV-Tat peptide enters the cell in endocytic vesicle, which travels along the cytoskeleton fibers and is released on the opposite side of the cell via exocytosis, effectively crossing the blood–brain barrier. If the HIV-Tat peptide is covalently attached to the outer surface of a VLP loaded with a drug of interest, the VLP is guided across the blood–brain barrier where it would release its cargo. Proof of principle experiments show that bacteriophage P22 VLP loaded with pain-relieving neuropeptides and displaying the HIV-Tat peptide of its outer surface successfully crosses in vitro and

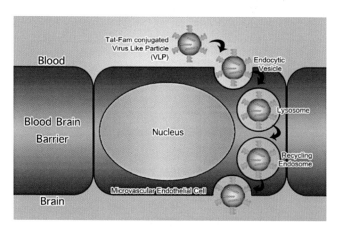

FIGURE 6.11 **Transcytosis of Tat conjugated viruslike particle (VLP) across endothelial cell binding the wall of blood vessels.** The Tat peptide from HIV is known to get across endothelial cells building the walls of blood vessels, effectively crossing the brain–blood barrier. Fluorescently labeled Tat peptide (Tat-FAM) was conjugated to the CP of bacteriophage P22 and assembled into a VLP encapsulating pain-relieving snail peptide fused to P22 scaffold protein. The resulting particle was easily traceable and was able to cross the model for blood–brain barrier, offering proof of principle evidence for future use of the approach. *Reprinted with permission from Anand P, O'Neil A, Lin E, Douglas T, Holford M. Tailored delivery of analgesic ziconotide across a blood brain barrier model using viral nanocontainers.* Sci Rep 2015;5:12497. https://doi.org/10.1038/srep12497; CCBY, http://creativecommons.org/licenses/by/4.0/.

in vivo models of the blood–brain barrier. The cargo neuropeptide was expressed and purified as a fusion to truncated P22 scaffold SP, which naturally interacts with the inner surface of the capsid. Fluorescently labeled HIV-tat peptide (Tat-FAM) was chemically conjugated to a mutant P22 CP in which a methionine has been replaced with cysteine, thus creating a traceable nanodelivery vehicle.

Controlled cargo delivery is gaining more attention as a possible avenue to ensure safety of VLPs/VNPs. Limited number of ideas has been explored so far. Based on the observation that the capsid of some viruses (HIV, hepatitis B and C, among others) contains disulfide bonds, it has been proposed that engineering disulfide bonds might be used as an avenue to increase VLP stability as well as to potentially control VLP opening and unloading. The latter is based on the notion that the cytoplasm is a reducing environment, whereas extracellular space is oxidizing, thus VLPs with disulfide bonds are likely to be stable while circulating in the body and to easily open once inside the cell. The approach is most compatible with cell-free systems for protein expression, which allow straightforward postpurification manipulations under conditions with controlled redox state. Another strategy for design of VLPs with controlled release of their cargo is to use capsids modified with 5-Norbornene-2-carboxylic acid and Grubbs' catalyst to trigger ring-opening metathesis polymerization (ROMP) reaction under physiological conditions (Fig. 6.12). The strategy

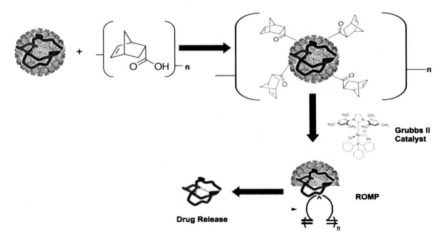

FIGURE 6.12 **Proof of principle design of viruslike particle (VLP)–based dissociative nanocontainer for controllable drug delivery.** Bacteriophage P22 VLP is conjugated to strained olefins that undergo a ring-opening polymerization reaction (ROMP) in the presence of Grubbs' II Catalyst, thus disrupting the integrity of the nanocontainer and releasing the encapsulated drug. *Image courtesy of Kelly P, Anand P, Uvaydov A, Chakravartula S, Sherpa C, Pires E, O'Neil A, Douglas T, Holford M. Developing a dissociative nanocontainer for peptide drug delivery.* Int J Environ Res Public Health 2015;**12**(10):12543; https://doi.org/10.3390/ijerph121012543; CCBY, http://creativecommons.org/licenses/by/4.0/.

has been successfully applied to bacteriophage P22 VLPs in proof of concept experiments in vitro.

Finding feasible solutions for in vivo VLP targeting, stability, and controlled release of the cargo will allow research efforts to focus on interactions of VLPs with the immune system and their intracellular whereabouts. It is generally thought that VLPs, just like regular viruses, are prone to neutralization by phagocytes, which would diminish their effectiveness and mount immune response. On the other hand, VLPs enter the cell via endocytosis and should be able to withstand the acidic environment of the endosome. The broad spectrum of potential applications of VLPs and VNPs as nanodelivery vehicles promises exciting developments in the near and distant future.

6.4.2 Viruslike Particles /Virus Nanoparticles as Nanobioreactors

VLPs and VNPs offer unique perspectives to enzyme catalysis and its commercial application for manufacturing large-volume bioproducts: (1) they can provide microenvironments closely resembling the natural compartments, where enzymes work in the cell; (2) they increase enzyme stability and have the potential to improve the cost-effectiveness of manufacturing, and (3) VNP-based nanoreactors can propagate themselves, thus cutting manufacturing cost by bypassing costly enzyme purifications steps. The basic principles of the design of VLPs/VNPs nanoreactors are the same as the ones described above for the design of a nanodelivery vehicle loaded with or displaying proteins; however, additional constraints apply. Since enzymes are large protein molecules, steric interference needs to be taken into account. Many enzymes are multisubunit entities and the VLP/VNP to be engineered into a nanobioreactor needs to be versatile, allowing a variety of modifications to accommodate two or more polypetide chains. If enzymes are to be packaged in the internal cavity of VLPs, the substrate(s) and product(s) of the enzyme reaction should be able to shuttle freely across the proteinaceous shell. On the other hand, only enzyme molecules that retain activity following necessary modifications can be utilized in nanobioreactors. Conceptually, all of these challenges are very similar to the challenges of in vitro expression of active enzymes and the ones posed by VLP/VNP production and modification, thus designing functional bioreactors might come down to availability of a diverse set of VLPs/VNPs that can accommodate constraints driven by enzymes with different properties. Detailed understanding of the biochemistry of enzyme reactions would contribute greatly to rational design. Proof of principle experiments have demonstrated that VLPs/VNPs can be successfully engineered into nanobioreactors supporting enzyme functions with a variable impact on activity, while enzyme stability is generally increased. Enzyme encapsulation was successfully achieved passively (by incubating CCMV CP and horseradish peroxidase together under conditions promoting capsid assembly) and actively (using genetic fusions to SPs, charged peptides, nucleic acid tags, among others).

An elegant series of experiments utilizing bacteriophage P22 VLP to design nanobioreactors with increasing complexity demonstrates very well the challenges and the rewards of the field. P22 is a *Salmonella typhimurium* dsDNA phage, whose capsid is built from 420 copies of a single CP. The capsid assembly follows the conventional mechanism for dsDNA viruses. Initially, a procapsid particle is assembled with the help of a SP. CP and SPs interact with each other noncovalently, and it has been shown that the C-terminal portion of the SP is sufficient to sustain the interaction. Under normal circumstances, once the procapsid is formed, the SP molecules are gradually released and recycled for assembly of other procapsids and the DNA genome is packaged, resulting in a mature capsid. The mature capsid is slightly larger (64 nm diameter) and more angular than the procapsid (58 nm diameter). P22 VLPs can be produced by coexpression of the CP and SP, resulting in a procapsid-like VLP, which transitions to a mature capsid-like VLP by heating at 65°C for 10 min. Further heating at 75°C (20 min) triggers loss of CP subunits, effectively opening pores with 10 nm diameter while preserving the overall integrity of the particle, which is commonly called "Whiffle ball" (WB) capsid. Collectively, the knowledge of the basic biology of P22 capsid assembly allowed researchers to manufacture procapsids (58 nm diameter and 2 nm pores), mature capsids (64 nm diameter and 2 nm pores), and WB capsids (64 nm diameter and 10 nm pores), enabling them to address various questions in a well-controlled manner. Proof of principle experiments showed that fluorescent proteins (FPs) fused to the C-terminus of the SP can be encapsulated in P22 VLP via the SP/CP noncovalent interactions. Furthermore, the FP encapsulation did not impede the ability of the VLP to transition to a mature and WB capsid when heat was applied. The loaded VLP was found to be a rather dynamic entity. It was shown that the C-terminal part of the SP could go through the 2 nm pores of the procapsid VLP and be cleaved by thrombin provided in solution. When procapsid VLPs were heated and transitioned to WB capsids, the remaining FP-SP fusion protein was able to diffuse across the 10 nm pores. These experiments, as well as similar ones in other systems, demonstrated that VLP walls are permeable and it is realistic to think about them as possible nanobioreactors taking up substrates and releasing products, providing the existence of pores with appropriate sizes. In addition, it became clear that the utilized mechanism of encapsulation (fusion to SP) does not mean that the cargo is anchored to the wall of the capsid, strongly suggesting that soluble enzymes are likely to tolerate the encapsulation protocol. At the next step of the experimental series, the enzyme alcohol dehydrogenase D (*AdhD*, 32 kDa), from the hyperthermophile *Pyrococcus furiosus,* was encapsulated in P22 VLP using the SP genetic fusion approach. Both, CP and the fusion protein, were expressed simultaneously from a single plasmid and the self-assembled VLPs encapsulating AdhD-SP were purified by ultracentrifugation. It was shown that on average each VLP contained 250 molecules of AdhD-SP that constituted a local concentration of 7 mM or 300 g/L, which is thought to be close to the protein concentrations in the context of the living cell. The activity of the encapsulated enzyme appeared lower than the one of the free molecule in a manner independent of the type of the VLP (procapsid, mature, or WB capsid) and at least in part consistent with effects of overcrowding, also observed at high enzyme concentrations in a test tube. Again, the abundance of prior knowledge regarding the thermostability of the enzyme and its ability to tolerate fusions, as well as the availability of a straightforward method to monitor the conversion of substrate, were key for the success of the experiment.

Many enzymes in living organisms are complex multisubunit entities, and thus the next proof of principle experiment in this sequence demonstrated that a two-subunit hydrogenase can be efficiently encapsulated in P22 VLPs and eventually used as a renewable catalyst of hydrogen fuel manufacturing (Fig. 6.13). Both subunits were cloned as SP-fusions coded on the same plasmid and their expression was induced by IPTG. The P22 CP was encoded on a second plasmid and its expression induced with arabinose. Simultaneous expression of all three proteins resulted in encapsulation of both hydrogenase subunits; however, the enzyme activity was much lower than the one of the free enzyme. Since both hydrogenase subunits are encoded by an operon with a total of six genes, it was reasoned that the other four proteins may help the hydrogenase mature to active form. To account for this possibility, the induction of hydrogenase subunits and the P22 CP was decoupled by inducing the CP with a 4 h delay. The resulting encapsulated mature hydrogenase demonstrated 100-fold higher activity than the free enzyme along with increased thermal and proteolytic stability.

Continuing the progression from simple to more complex, the next body of work with the P22 system aimed to encapsulate three enzymes catalyzing a cascade of coupled reactions (Fig. 6.14). Furthermore, the stoichiometry of the multienzyme complex was rather challenging as one enzyme was a monomer, one a dimer, and one a tetramer. The three enzymes were genetically fused as consecutive blocks with one SP tag and flexible connecting linkers, resulting in a giant fusion protein (160 kDa) coexpressed together with the CP from one plasmid. Although VLPs encapsulating all three enzymes in active form were identified, low yields of the fusion prevented in-depth studies. Nevertheless, new knowledge was acquired regarding the significance of the order of individual proteins in the fusion polyprotein, channeling of molecules between colocalizing enzymes,

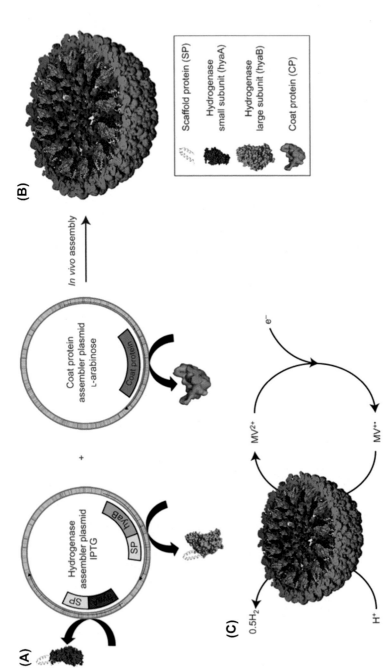

FIGURE 6.13　**Proof of principle design of hydrogen fuel nanoreactor.** P22 viruslike particle (VLP) was engineered to encapsulate two-subunit hydrogenase. The nanoreactor was supplied with a methyl viologen (MV+) as an electron transfer mediator and dithionite as source for electrons. Hydrogen production was monitored via gas chromatography. Both subunits of the hydrogenase (in red and green) were expressed as individual fusions to P22 scaffold protein (SP; in yellow) coded on the same plasmid under the control of IPTG-inducible promoters. The coat protein was expressed in parallel from a separate plasmid under the control of an arabinose-inducible promoter. *Reprinted from Jordan PC, Patterson DP, Saboda KN, Edwards EJ, Miettinen HM, Basu G, Thielges MC, Douglas T. Self-assembling biomolecular catalysts for hydrogen production. Nat Chem 2016;8(2):179–85, with permission from Nature Publishing Group.*

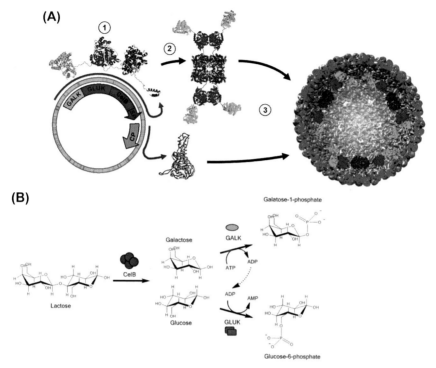

FIGURE 6.14   **Proof of principle design of P22 viruslike particle (VLP) metabolon.** P22 VLP was engineered to encapsulate three enzymes from the same metabolic pathway: CelB (in red), GALK (in green) and GLUK (in blue). All enzymes were expressed as one polyprotein fused to the P22 scaffold protein (SP) domain coded by a plasmid also expressing the P22 capsid protein (CP) from a different open reading frame (A). Each enzyme catalyzes an individual reaction from the same metabolic pathway (B). *Reprinted from Patterson DP, Schwarz B, Waters RS, Gedeon T, Douglas T. Encapsulation of an enzyme cascade within the bacteriophage P22 virus-like particle.* ACS Chem Biol 2014;9(2):359–65, *with permission Copyright (2012), American Chemical Society.*

the biochemistry of the individual enzymes, and coupling within different enzyme pairs. The described sequence of research endeavors reflects well the challenges and rewards scientist experience pushing forward. Complexity of the living matter is often a cause for pause in research advancement, as well as a powerful motivator for development of new approaches and technology.

An alternative approach to assemble multienzyme nanoreactors was developed based on CCMV VLPs. Nucleic acid tags were used to encapsulate enzymes of the same metabolic pathway on the positively charged inner surface of the VLP. It was shown that low-molecular-weight substrates and products, such as glucose, hydrogen peroxide, ATP/ADP, can easily diffuse across the capsid wall. In one example, glucose was converted to ribulose-5-phosphate in a cascade of three enzyme reactions. The substrate, glucose, entered the nanoreactor VLP via diffusion and was subjected to two coupled enzyme reactions. The product of the second reaction diffused outside the VLP, where it was utilized as a substrate of the third enzyme from the cascade catalyzing the conversion to ribulose-5-phosphate.

VNP-based bionanoreactors utilize phage display methodology to express an enzyme of interest on the surface of a self-propagating virus. Selected phages are propagated in the host bacterium and expected to be directly added in the manufacturing processes, saving cost and resources for enzyme purification. For example, α-amylase was displayed on the surface of bacteriophage T7 as a fusion to its CP, gp10. It was shown that the displayed enzyme has comparable performance and stability to the commercially used purified protein. A tandem fusion of α-amylase and xylanase A was also tested in the context of T7 phage display revealing that the approach is not optimal since the activity of the enzyme closer to the capsid surface is somewhat inhibited. Both enzymes are of significant economic importance with applications in ethanol production, textile, and paper industries.

6.4.3 Virus-like Particles /Virus Nanoparticles as Imaging Probes

Viruses and their VLPs are of interest for developing a new generation of imaging agents for biomedical applications due to their natural affinity for certain biological structures, their cargo-carrying capacity and potential for modifications directing targeted delivery, and unique way of influencing the properties of traditional imaging probes resulting in improvement of image quality and resolution. Ideas to use VNPs and VLPs as imaging agents have been explored for intravital (live) imaging of vasculature using fluorescent dyes in fluorescence tomography, PET and MRI imaging, all with tremendous diagnostic value.

It has been known that CPMV has a natural propensity for endothelial cells and it is readily internalized by them, although the specific mechanisms are not understood. CPMV is a plant virus entering its host through mechanical injury and spreading systemically. The virus does not execute productive infection in animals despite the fact that it gets engulfed by endothelial cells. Inside the cell, CPMV is found in endocytic vesicles colocalizing with lysosomal and trans-Golgi markers. In chicken embryos, fluorescently labeled CPMV persists for up to 2 days, which is consistent with the stability of the viral capsid at low pH. Each VLP contains 100+ fluorophore molecules providing excellent quality of staining. Since the fluorophores can be attached only in a limited number of specific positions and the cavity of the VLP is large enough for the fluorescent molecules to not interfere with each other, photo bleaching is not an issue. Sequential injections with CPMV carrying different fluorophores allows for tracking vasculature development over time, which could be a useful tool for evaluation of the progression and/or inhibition of tumor vascularization. Biodistribution studies show that CPMV is also found in liver and spleen. The addition of PEG molecules on the outside of the VLP abolishes its ability to localize to endothelial cells and diminishes retention by liver and spleen. Fluorescently labeled canine parvovirus (CPV) has a propensity to stain tissues expressing the transferrin receptor, which happens to be its natural receptor in dogs. CPV can bind and enter human cells utilizing the human transferrin receptor; however, its infection is somewhat inefficient, thus presenting an opportunity to develop an imaging agent based on the CPV VLP. Transferrin receptor and folic acid receptor are found to be overexpressed in many tumors, allowing transferrin and folic acid to be utilized as targeting agents for imaging or drug delivery purposes. VLP modifications attaching tissue/organ-specific targeting peptides identified by in vivo biopanning, have the potential to convert any VLP into an imaging probe.

The ability of VLPs to support multivalent encapsulation/display of ligands is an attractive feature from the perspective of development of paramagnetic MRI contrast agents. Currently, medicine uses gadolinium for this purpose, most frequently injected as a salt in a complex with diethylenetriaminepentacetate (DTPA). Free Gd^{3+} is toxic, thus any VLP that is to be used to deliver a contrast agent should bind the ion with high affinity. VLPs with natural propensity for binding metal ions such as CPMV and Qβ have been explored as carriers of a contrast agent. P22 and MS2 VLPs have been chemically modified to encapsulate gadolinium. VLPs are very good candidates for development of contrast agents since they allow multiple Gd^{3+} ions to be immobilized on the inner and/or outer surface, and their size allows for modulation of the relaxation times of protons in magnetic field, thus improving the strength and the sharpness of the signal. The capsids with WB configuration are of special interest for development of MRI contrast agents since they are large, tumble slow, and thus modulate proton relaxation better. In addition, the pores of the WB capsids allow for diffusion of water molecules, which has the potential to influence the quality of the signal. Ideas for alternative contrast agents explore the possible application of hyperpolarized xenon atoms trapped into VLPs with the help of cryptophane cages, a complex organic entity. Bacteriophage MS2 VLPs have been modified to encapsulate such cages. Xenon is nontoxic and can be introduced in the body via inhalation. When placed in a magnetic field, trapped xenon atoms behave differently than free ones, thus creating an imaging signal. The strength of the xenon signal significantly exceeds the one of the conventional MRI approach and would potentially allow for higher resolution scans to be performed in very short times.

Similar approaches are taken for development of PET imaging probes (Fig. 6.15). In one example the MS2 capsid protein has been genetically engineered to allow chemical modification with maleimide-DOTA, which is capable of coordinating ^{64}Cu thus rendering the particle radioactive. Whole animal dynamic PET imaging of mice with transplanted tumors was performed to assess the body distribution of the radioactive particles and the quality of the imaging. In a subset of experiments, PEG chains were attached on the outer surface for the purpose of reducing the VLP charge and immunogenicity, and were shown not to interfere significantly with the imaging process. It is expected that targeted VLP delivery of PET imaging agents will allow for high resolution imaging with minimal irradiation of healthy tissues.

Developing virus-based imaging agents is a young field that is poised for rapid growth once critical mass of knowledge and experimental data are accumulated for the stability, distribution, and toxicity of various VLPs and their ability to withstand various conjugation/modification chemistries. Immune response to viral particles could be of potential concern; however, the issue is common to all virus-based biomedical applications, thus allowing significant research efforts to be applied toward finding solutions to bypass the challenge.

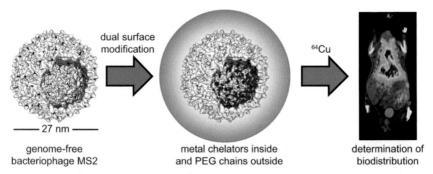

FIGURE 6.15 **Dual-surface modification strategy for generation of bacteriophage MS2 viruslike particle (VLP) for PET biomedical imaging.** The MS2 capsid protein has been engineered to include a p-aminophenylalanine mutation (T19paF; in blue) on the VLP outer surface and a cysteine mutation (N87C, in red) on the VLP inner surface, each allowing for specific chemical modification. Maleimide-DOTA was attached to the interior cysteines to allow ^{64}Cu binding, thus rendering the particle radioactive. PEG chains were attached on the outer surface for the purpose of reducing the VLP charge and immunogenicity. Whole animal dynamic PET imaging of mice with transplanted tumors was performed to assess the body distribution of the radioactive particles and the quality of the imaging. *Reprinted from Farkas ME, Aanei IL, Behrens CR, Tong GJ, Murphy ST, O'Neil JP, Francis MB. PET imaging and biodistribution of chemically modified bacteriophage MS2. Mol Pharm 2013;10(1):69–76, with permission Copyright (2013), American Chemical Society.*

6.4.4 Virus-like Particles /Virus Nanoparticles as a Platform for Development of Nanomaterials and Nanodevices

Due to their size, defined structure, and diversity, viruses and their VLPs present useful renewable resources for development of nanomaterials and nanodevices. Their accessibility, cost-effective production, and the level of homogeneity make them superior to any synthetic nanoparticle currently in manufacturing or experimentation. The enormous potential for modifications by genetic and chemical approaches, as well as the huge number of viruses in nature with a broad range of sizes, symmetry, and physical/chemical properties, creates an extensive array of options to be explored while seeking solutions to practical problems. Furthermore, viruses infecting organisms living in extreme environments are of particular interest due to their extraordinary stability at high temperature, various pH, salt concentration, etc. VNPs and VLPs of some viruses exhibit surprising stability in organic solvents, allowing for extensive chemical modifications. Tailed bacteriophages offer an advantageous combination of naturally connected icosahedral and tubular shape in one particle with the possibility for each one of them to be functionalized individually, thus providing opportunities for modeling transitions between different states. Due to their protein composition, VNP and VLP-based nanomaterials are less likely to be toxic that synthetic ones.

Metallization is a key process in the design of virus-based nanomaterials. Metal ions and inorganic materials in general have been successfully added on the internal and external surfaces of different viruses, resulting in metal nanoparticles, nanotubes, and wires with application in electronics and sensing. Two principal approaches are utilized toward VNP and VLP metallization: depositing metal ions on the surface of viral capsids (Fig. 6.16A and B) and using metal nanoparticles as a platform to assemble VLPs (Fig. 6.16C and D). The internal and/or external surfaces of viral capsids are naturally charged. Electrostatic interactions between capsid proteins and nucleic acids are essential for genome packaging and maintenance of the integrity of the virion. On the other hand, the charged outer surface facilitates virus adsorption on the host cell via interactions with polysaccharides, glycoproteins, and charged proteins. Some capsids, for example the CPMV capsid, naturally coordinate Ca^{2+} under physiological conditions, which can be replaced with other metal ions in vitro. Due to the presence of a large number of metal-coordinating sites, metal ions or metal clusters bound to capsid proteins can serve as nucleation sites and drive the growth of metal nanoparticles with precise shape and size limited by the size of the capsid cavity or the surface area of the VLP. Premanufactured nanoparticles with charged surfaces can be used as platforms for capsid assembly directly or after modification. Theoretically, both principal strategies can be combined to achieve

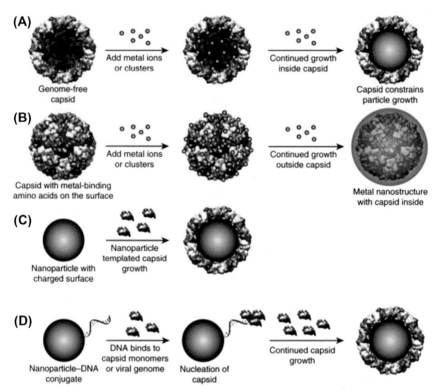

FIGURE 6.16 **Conceptual application of viruslike particles (VLPs) toward formation of inorganic (metal-containing) nanostructures.** Metal ions and/or clusters can bind electrostatically or in a sequence-dependent fashion to the inner (A) or outer (B) surface of VLPs utilizing naturally charged capsid proteins or engineered fusions to metal-binding proteins. High local concentration of metal materials can further serve as a nucleation point for particle growth. Nonmodified (C) or modified (D) nanoparticles can be used as platforms for capsid assembly and the resulting VLPs used as a delivery vehicle of the metal nanoparticle, for example, in phototherapy. The same approaches are applicable for helical/filamentous capsids such as TMV or bacteriophage M13. *Reprinted from Stephanopoulos N, Francis MB. Making new materials from viral capsids. In: Matyjaszewski K, Möller M, editors.* Polymer Science: A Comprehensive Reference. *Amsterdam: Elsevier; 2012. p. 247–66, with permission from Elsevier.*

an inner core and outer surface with different chemical nature. Viral capsids with tubular shape are proposed to give rise to nanowires when subjected to metallization. Furthermore, genetic or chemical modifications of the outer surface of a capsid shell with any shape are being used to attach a metallized capsid on desired surfaces in a configuration of interest, taking us closer to potential materials with application in nanoelectronics. An interesting mechanism of pH-controlled capsid transitions has been described for red clover necrotic mosaic virus (RCNMV). In acidic environment, the capsid is very compact, whereas when pH

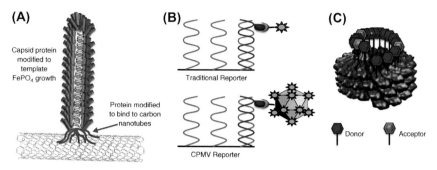

Current Opinion in Chemical Biology

FIGURE 6.17   **Example applications of viral particles in nanomaterials.** (A) M13 phage designed to bind to carbon nanotubes and nucleate metal ions to create a nanobattery cathode. (B) Cowpea chlorotic mottle virus (CPMV) chemically modified with multiple fluorophores to be used as a reporter in DNA microarrays. (C) Tobacco mosaic virus (TMV) displaying donor and acceptor chromophores to function as a light harvesting system. *Reprinted from Koudelka KJ, Manchester M. Chemically modified viruses: principles and applications.* Curr Opin Chem Biol 2010;**14**(6):810–7, *with permission from Elsevier.*

starts increasing, the shell swells, resulting in pore opening that allows small molecules to diffuse inside the capsid. Positively charged small molecules bind electrostatically to the RNA genome and are retained inside the virion. Lowering the pH transitions the capsids back to compact shape with no pores, thus trapping a defined amount of small molecules inside the viral particle.

Bacteriophage M13 and phage display are of particular interest in terms of developing new materials. Biopanning has been used to select for peptides binding to metals, metal oxides, semiconductors, quantum dots, and other inorganic materials. A true testament for the power of phage display is the selection of peptides differentially binding to alternative crystal forms of the same semiconductor material. The versatility of M13 in terms of display options allows researchers to combine multiple display approaches on one particle, which is especially beneficial for creating compact size, environmentally friendly lithium batteries with high capacity. In a proof of principle experiment, M13 was attached to carbon nanotubes utilizing pIII minor CPs, whereas the major CP (pVIII) was genetically modified to coordinate iron (alpha-$FePO_4$), collectively creating a bacteriophage-based cathode (Fig. 6.17). Cathode pairing with lithium metal foil anode resulted in a nanobattery. A fully bacteriophage-based battery was assembled from the above cathode, cobalt-metallized phage as an anode, and silver viral nanowires all deposited on the same surface with creative spatial positioning approaches.

Due to their elongated shape and homogeneity, filamentous bacteriophages are being studied as liquid crystalline model systems and options for controlled ordering are being explored. Applying soft lithography techniques, researchers were able to immobilize anti-M13 antibodies on gold surface. Subsequent incubation with phage solutions produced a highly ordered structure, which in the near future could become a useful element of nanoelectronic devices. M13 particles displaying ZnS-binding peptides have been reported to self-assemble into periodically ordered film structures of ZnS nanoparticles. Similar findings have been reported for magnetic nanoparticles based on coordination of iron or cobalt, as well as semiconductor nanomaterials based on coordinated gallium nitride. M13 and other viruses displaying specific peptides are also being used as scaffolds for tissue engineering (details are described in Chapter 5 as part of applications of phage display). Bacteriophage Pf1 isolated from *Pseudomonas aeruginosa* is used as a sample alignment tool in NMR spectroscopy to facilitate measurements of dipolar coupling interactions in protein and nucleic acid samples.

Phage display–selected peptides binding specific ligands are considered a valuable tool for developing nanosensors detecting chemically diverse components ranging from explosives to environmental pollutants and pathogen identification markers. It is hoped that the combination of the sensitivity of phage display and the compactness of future nanoelectronic devices will result in the design of cost-effective portable sensing equipment with broad capabilities for field research and biomedical applications. Sensing systems based on optical, nanomechanical, and electrochemical principles have been reported. For example, ELISA and microarray detection (Fig. 6.17) capabilities can be improved by replacing conventional fluorescently labeled reagents with fluorescently labeled VLPs, which coordinate a large number of fluorophores allowing for signal amplification. Fluorescently labeled viruses can be used as probes in FACS (Fluorescence-activated cell sorting) assay to identify cells with desired properties. An application of this approach for identification of phage absorption deficient bacteria in milk fermentation industry is described in Chapter 7. Sensing probes based on quartz crystal microbalances and surface plasmon resonance rely on changes in accumulating mass on the sensing surface that are translated into measurable signal due to sequential changes of the mechanical or optical properties of the platform. Such probes tend to have very high sensitivity and have the potential to influence the field of sensing similarly, to the way surface plasmon resonance influenced binding studies and single-molecule techniques are changing our understanding of biochemical processes initially studied only by steady-state equilibrium conditions.

Viruses are an attractive option for design of light energy harvesting and storage devices employing FRET (Förster resonance energy transfer). Fig. 6.17 depicts TMV modified with donor and acceptor chromophores arranged to harvest broad-spectrum light energy. The donor chromophore absorbs light energy from the environment, becomes excited, and releases the harvested energy as photons with particular wavelength, which are captured by the acceptor chromophore. If the acceptor chromophore is coupled to an appropriate chemical reaction, the harvested light energy can be converted into chemical energy in a similar manner to photosynthesis. Successful FRET energy transfer has been demonstrated between fluorescent dyes encapsulated in bacteriophage MS2 and chemically attached porphyrins on the outside surface of the capsid. Light emitted by the fluorescent dye travels through the 2 nm capsid wall to excite the porphyrins. The electron transport capacity of the system was registered by reduction of methyl viologen. The highly regular and repetitive structure of viral capsids creates opportunities for optimal spacing of chromophores that prevents photobleaching and ensures efficient energy transfer. The list of applications of VLPs and VNPs in nanomaterials is exponentially growing along with the number of approaches and viruses employed. The examples briefly described above should be viewed as an attempt to provide a glimpse of the big picture of a fast developing field, rather than presentation of its depth and breath. Additional research and biomedical applications related to phage display are discussed in Chapter 5.

Further Reading

1. Brasch M, Putri RM, de Ruiter MV, Luque D, Koay MS, Caston JR, Cornelissen JJ. Assembling enzymatic cascade pathways inside virus-based nanocages using dual-tasking nucleic acid tags. *J Am Chem Soc* 2017;**139**(4):1512–9.
2. Cardinale D, Carette N, Michon T. Virus scaffolds as enzyme nano-carriers. *Trends Biotechnol* 2012;**30**(7):369–76.
3. Hetherington CL, Moffitt JR, Jardine PJ, Bustamante C. 4.22 viral DNA packaging motors A2. In: Egelman, Edward H, editors. *Comprehensive Biophysics*. Amsterdam: Elsevier; 2012. p. 420–46.
4. Koch C, Wabbel K, Eber FJ, Krolla-Sidenstein P, Azucena C, Gliemann H, Eiben S, Geiger F, Wege C. Modified TMV particles as beneficial scaffolds to present sensor enzymes. *Front Plant Sci* 2015;**6**:1137.
5. Koudelka KJ, Pitek AS, Manchester M, Steinmetz NF. Virus-based nanoparticles as versatile nanomachines. *Annu Rev Virol* 2015;**2**(1):379–401.
6. Rakonjac J, Bennett NJ, Spagnuolo J, Gagic D, Russel M. Filamentous bacteriophage: biology, phage display and nanotechnology applications. *Curr Issues Mol Biol* 2011;**13**(2):51–76.
7. Rohovie MJ, Nagasawa M, Swartz JR. Virus-like particles: next-generation nanoparticles for targeted therapeutic delivery. *Bioeng Transl Med* 2017;**2**(1):43–57.
8. Tang J, Johnson JM, Dryden KA, Young MJ, Zlotnick A, Johnson JE. The role of subunit hinges and molecular "switches" in the control of viral capsid polymorphism. *J Struct Biol* 2006;**154**(1):59–67.
9. Wen AM, Steinmetz NF. Design of virus-based nanomaterials for medicine, biotechnology, and energy. *Chem Soc Rev* 2016;**45**(15):4074–126.
10. Young M, Willits D, Uchida M, Douglas T. Plant viruses as biotemplates for materials and their use in nanotechnology. *Annu Rev Phytopathol* 2008;**46**:361–84.

Virus-Driven Biocontrol Approaches

7.1 PRINCIPLES OF BIOCONTROL

Biocontrol is broadly defined as an approach or strategy to control pest species negatively influencing human activities such as agriculture, health, food production, waste management, etc. In a way, the efforts of medicine to control infectious diseases in humans present a form of biocontrol, although the term is rarely used in that context. Pest species are taxonomically diverse and range from microorganisms to insects, weeds, and even mammals. Biocontrol is an integral component of integrated pest management or IPM, and is frequently discussed in conjunction with organic farming and minimizing/eliminating pesticide use. Often various biological agents used in biocontrol are also called biopesticides, although the precise definition of biopesticides varies depending on the agency and may include living organisms or products originally isolated from them. As an approach, biocontrol relies on detailed knowledge of interspecies or host/virus relationships in the context of the specific ecosystem. Most commonly, biocontrol takes advantage of man-managed natural examples of predation, herbivory, and parasitism applied toward reduction of the population size of the target pest. Human management falls in three major categories: (1) classical biocontrol; (2) conservation biocontrol; and (3) augmentative biocontrol.

Classical biocontrol is most frequently applied toward control and eradication of invasive species. These are organisms new to a particular environment, which propagate without any restrictions due to lack of natural enemies, thus reshaping the ecological landscape of the ecosystem. Trade and transportation are common means of introduction of invasive species. Classical biocontrol methods identify the exotic species, study its relationships in the ecosystem/geographical location of origin, and select a natural enemy to be imported in the infested area as a counterbalance measure. The imported predator is subject to quarantine and extensive

195

evaluation before being released in the target ecosystem to avoid trans-fer of unwanted "hitchhikers." The classical approach is most frequently applied toward invasive insects, although theoretically it can be applied to a species of any taxonomic group, including mammals. Programs tar-geting native pest species tend to be less successful than the ones aiming at newly introduced invasive ones. Either way, continuous monitoring is a key element of the approach and most of the time classical biocontrol is successful. It is also cost-effective since it requires only initial investment and can self-maintain over long periods of time.

Conservation biocontrol is an approach reinforcing preservation of a diversity of ecosystems, so that the number of pest species is kept under control by natural mechanisms. Essentially, human efforts are focused on avoiding overuse of ecosystems so that economic activity does not dis-turb the already established balance of species. At minimum, conservation biocontrol calls for limited use of a broad range of pesticides, which can impact species beneficial for pest control.

Augmentative biocontrol is exercised by habitat alterations or increasing numbers of natural enemies. For example, planting flowering plants attract-ing bees is helpful to increase bee pollination. Similarly, planting species that provide shelter of the natural enemy of the target pest effectively contrib-utes to the increase of the number of predating individuals. In contrast, crop alternation limits the overgrowth of common pests. More involved mea-sures include controlled breeding and release of natural agents attacking the target pest, thus supplementing the action of the agents already pres-ent in the ecosystem. The release can be done in limited numbers during a critical target period, commonly referred to as inoculative release, or in very large numbers, commonly referred to as inundative release. Inoculative release aims to prevent expected infestation, whereas the inundative release tends to be used as a "clean up" process of an already apparent problem. Augmentative biocontrol approaches require continuous investments and a deep understanding of the complexity of the ecosystems they are applied to.

The biggest challenge in application of biocontrol approaches is the reliable prediction of long-term outcomes. In part, that is due to the enor-mous complexity of ecosystems and the adaptability of living matter. On the other hand, limited control over the course of events allows small inci-dents to result in a significant impact. Legislation related to biological con-trol is highly variable throughout the world and often at different stages of development depending on the pest issues each country is facing. As a general trend, predation-based approaches gain public support easier than pathogen-based approaches. Biocontrol measures targeting insects and weeds are better received as opposed to biocontrol approaches target-ing mammals. For example, at the very beginning of this century, there were more than 5000 introductions of natural enemies of insects and mites that were and remain mostly out of public discussion. In contrast, the very few proposed releases of natural agents targeting vertebrates have been subject to intense opposition. Safety concerns present only one side of the problem and will be eventually resolved with extensive testing, extended time, and funding. Another aspect of opposition is social perceptions of pest versus nonpest organism. Insects are considered more of a pest than a friend, and thus the perception of biological control targeting insects is more likely to be positive than negative. In contrast, vertebrates are closer to humans evolutionary, many of them are considered pets or livestock, and thus trigger concerns related to animal welfare and emotional conflicts driven by the close association with them. The third aspect of opposition stems from the negative reaction against genetically modified organism. Most likely the importance of biocontrol approaches will continue to rise as the society is becoming more conscientious about the environment and sustainability, as well as various genetic manipulations are making their way into medicine. The *One Health* philosophy is urging for integration of global, national, and local efforts to promote and improve the health of humans, animals, and the environment as one whole complex entity. Coordinated efforts focusing on solving problems from multiple perspectives have proven to be an effective strategy in many settings and are certainly essential for biocontrol approaches.

Since viruses infect organisms of all domains of life and most have narrow host range, they are good candidates as biocontrol agents, assuming enough specifics of host/virus interactions are understood and methods for laboratory growth of the viruses of interest exist. Among the biggest concerns in virus-based biocontrol is unexpected vector-mediated transfer and host adaptability leading to emergence of virus resistance.

7.2 APPLICATIONS OF VIRUSES TO CONTROL THE POPULATION SIZE OF VERTEBRATE INVASIVE SPECIES

Although invasive species are a threat to natural ecosystems worldwide, the severity of the problem is very evident in Australia. Due to its natural uniqueness and colonization history, the continent has "acquired" at least 80 species of nonindigenous vertebrates and approximately one-third of them are currently with a pest standing with established large populations in the wild. Correspondingly, Australia as a country has rich traditions in research and development of biocontrol approaches, as well as long-term studies on their economic effectiveness. Two of the three most accepted virus-based biocontrol programs targeting vertebrates have been developed in Australia (Fig. 7.1). Currently, Australia offers its citizens regular updates on various efforts to control invasive species and toolkits how to contribute to them (http://www.pestsmart.org.au/).

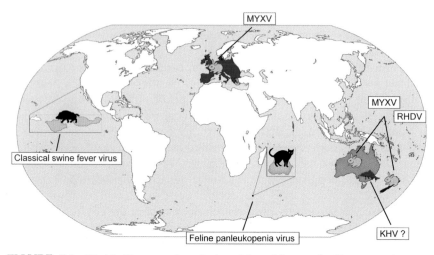

FIGURE 7.1 **Worldwide examples of virus-driven biocontrol efforts targeting vertebrate invasive species.** Successful efforts include the application of (1) Feline panleukopenia virus to eradicate cats on Marion Island off the coast of South Africa (magnified on the map in grey) and (2) Myxoma virus (MYXV) in Australia and Europe and RHDV rabbit haemorrhagic disease virus (RHDV) in Australia and New Zealand to control rabbit populations. Efforts to use koi herpesvirus (KHV) for carp control are currently in progress in Australia. Attempts to use classical swine fever virus to control the population of wild boar on the Channel Islands in the United States (the islands are magnified, with the two islands affected, Santa Cruz and Santa Rosa, shown in grey) were unsuccessful. *Image courtesy Di Giallonardo F, Holmes EC. Exploring host-pathogen interactions through biological control. PLoS Pathog 2015;**11**(6):e1004865.* https://doi.org/10.1371/journal.ppat.1004865, *CC-BY,* https://creativecommons.org/licenses/by/4.0/.

7.2.1 The Myxoma Versus Rabbits Saga

The efforts of controlling the population of European rabbits in Australia is a classical example of the challenges invasive species pose and the ups and downs of biocontrol approaches to limit their impact on the environment. European rabbits arrived in Australia with the first European settlers in the late 18th century. They were viewed as a symbol of home, a useful food resource, and were meant to be raised only in captivity. That worked very well until very limited number of European rabbits were released in the wild for the purpose of hunting. The Australian climate, the specifics of rabbit reproduction (each female rabbit can produce up to 35 bunnies each year), and the absence of natural predators favored exponential increase of the local population of free-roaming pests, followed quickly by its rapid spread all across the continent. The effect on the environment was devastating as rabbits were grazing on every plant they could possibly get access to and thus actively competing not only with wild animals with a similar ecological profile but also with farm animals (by limiting the amount

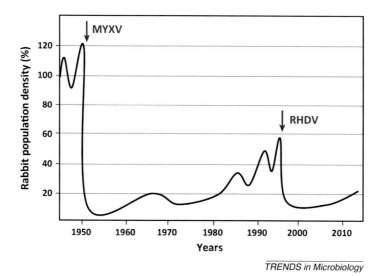

FIGURE 7.2 **Dynamics of the rabbit population in Australia.** The arrows indicate the onset of biocontrol efforts with myxoma virus (MYXV) in 1950 and rabbit hemorrhagic disease virus (RHDV) in 1995. The rabbit population of 1945 is set as a reference at 100%. *Reprinted from Di Giallonardo F, Holmes EC. Viral biocontrol: grand experiments in disease emergence and evolution.* Trends Microbiol 2015;23(2):83–90, *with permission from Elsevier.*

of grasslands) and with humans (by damaging vegetable crop). Initially, mainly physical methods targeting the pest rabbits were used. Rabbits were hunted, fences were built; however, the huge rabbit population barely got affected. When chemical poisoning failed to make a difference too, the search for biological means to control the pest began. Initially, various bacterial pathogens were considered but were not able to deliver significant results. In 1950, the myxoma virus (MYXV) escaped quarantined facilities after 15 years of research and infected the free-roaming rabbit population, quickly reducing the rabbit population from an estimated 600–100 million. Due to adaptation and rising resistance, the rabbit population fluctuated up and down but remained at levels significantly below the premyxoma ones (Fig. 7.2). Two flea vectors, European rabbit flea, *Spilopsyllus cuniculi*, and Spanish rabbit flea, *Xenopsylla cunicularis*, were bred in the lab and released in huge quantities to facilitate myxoma transmission. By the 1990s, the rabbit numbers bounced back up to 200 million, triggering interests in an alternative biocontrol agent: the rabbit hemorrhagic disease virus (RHDV), a calicivirus. RHDV was released from field-testing trials by accident and reduced the rabbit population by 90% within a year. Similar to myxoma, the pest became apparently resistant within a few years. The success of RHDV was somewhat variable, with the best results in areas with a drier climate.

Later it was discovered that a natural calicivirus with low pathogenicity is widespread in the rabbit population, most likely "vaccinating" the animals.

The MYXV is a large enveloped dsDNA virus from the Poxviridae family, with elaborate strategies for evading the immune system. Research on poxviruses was initially driven by the smallpox devastation (caused by the vaccinia virus, genus *Orthopoxvirus*) and later smallpox vaccination and eradication efforts. MYXV is a distant cousin of the vaccinia virus and belongs to the genus *Leporipoxvirus*. Two major strains have been isolated from nature (Brazilian and Californian), both infecting American rabbits (genus *Sylvilagus*), in which they cause a mild disease. In European rabbits (genus *Oryctolagus*), myxoma causes myxomatosis, a generalized disease with a fatal outcome involving fever and severe swelling in various parts of the body, especially in the eyes and genitals. The virus is transmitted by various animal vectors such as mosquitoes, fleas, and mites. Some strains can be transmitted from rabbit to rabbit.

RHDV is a calicivirus causing a severe and highly contagious disease resulting in fast liver necrosis and death. The virus itself is a naked icosahedral particle with (+)ssRNA genome. The virus is spread by direct contact, fomites, scavenging animals eating carcasses of dead infected animals, as well as mosquitos and fleas.

Biological control of rabbits in Australia is an ongoing battle as the release of each new biocontrol agent is counteracted by resistance and adaptation of the host, thus necessitating periodic interventions. Complexity of biocontrol applications is very well demonstrated by the attempted use of the myxoma approach in New Zealand and Europe after its well visible success in Australia. The same viral strains did not deliver significant reduction of local rabbit populations, most likely due to cold and wet weather that limits the propagation of mosquitos and sand flies. In the absence of a large number of vectors, the rate of transmission of myxoma cannot outpace the rate of rabbit propagation, thus rendering the approach ineffective. RHDV was illegally imported in New Zealand and worked with comparable success to Australia with a lethality rate of 84% after the initial applications. In Europe, MYXV was used privately for biocontrol a couple of years after its efficiency was evident in Australia. Unfortunately, the virus escaped and wiped out most of the rabbit population at the time. In later years, a highly virulent RHDV strain naturally appeared, however, with lesser impact, presumably due to cross-immunity. Ironically, today in Europe, where rabbits are bred as a delicatessen food source, the animals are vaccinated against both myxoma and RHD viruses to prevent losses. Spain even has a program to increase rabbit populations of several national parks in an attempt to increase the numbers of their natural predators whose populations are shrinking.

A recent study compared the evolution of myxoma and RHD viruses in Australia and Europe in parallel (Fig. 7.3), attempting to identify key principles to be applied to future biocontrol targeting mammals. It was discovered that the available data are consistent with virulence and transmission-based selection. The idea of virulence selection has been rationalized extensively postulating that viruses evolving to high virulence are bound to become extinct since they kill their hosts too fast, thus limiting their spread. In both continents, the currently circulating myxoma strains are milder than the ones originally released. In contrast, RHDV strains continue to be highly virulent. Furthermore, RHDV RNA was found in the livers of mice living in overlapping habitats without evidence of viral propagation or pathology. The analysis concluded that there is a strong selection for transmissibility, which may or may not be connected to diminishing virulence. Overall, despite the multiple ups and downs in the process of controlling rabbit population in Australia, the economic and ecological benefits are well recognized. For example, it has been estimated that 12 million Australian dollars have been spent over 8 years of research and development before RHDV was accidentally released in nature, whereas the net benefits were calculated to 106 million/year in agricultural productivity in the immediate years following the release.

An attempt to seek a more permanent solution to the rabbit problem was made with the development of genetically modified MYXVs expressing rabbit zona pellucida proteins involved in egg development and control of fertilization. Genetically engineered viruses were expected to work as an immunocontraception vaccine triggering autoimmune response

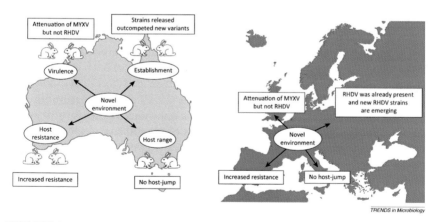

FIGURE 7.3 **Parallel evolution of myxoma virus (MYXV) and rabbit hemorrhagic disease virus (RHDV) in Australia and Europe.** Rabbits facing right represent MYXV infection, whereas rabbits facing left denote RHDV infection. In both continents, MYXV and RHDV spread rapidly. While MYXV attenuated, RHDV remained highly virulent. An increase in host resistance was observed for both viruses with no documented species jump beyond lagomorphs. *Reprinted from Di Giallonardo F, Holmes EC. Viral biocontrol: grand experiments in disease emergence and evolution.* Trends Microbiol 2015;**23**(2):83–90, *with permission from Elsevier.*

against the above proteins and conferring infertility. Although the lab research outcomes were very promising, the field testing demonstrated that the immunocontraceptive myxoma strain was outcompeted by its natural counterparts and the approach did not result in any large-scale releases of new biocontrol agents. Alternative viral agents such as herpesvirus isolated from domesticated rabbits in Canada, rabbit fibroma virus, and rabbit vesivirus are being explored as possible future tools to control the pest.

7.2.2 The Marion Island Program for Eradication of Feral Cats

Another big invasive species for Australia are feral cats that are currently threatening to disturb the ecological balance by eating a large number of local species, whose abundance is diminishing rapidly. A similar problem was resolved successfully on Marion Island, a subantarctic island in the Indian Ocean and part of the territory of South Africa, where a feline parvovirus, Feline Panleukopenia Virus (FPLV), was released as a biocontrol agent in the 1950s. The virus reduced the cat population drastically and the remaining animals were gradually eliminated using traps and hunting. Subsequently, the bird populations have bounced back and the ecological disaster has been reversed. Interestingly, the feline parvovirus did not disappear; instead, it is now infecting various local carnivores. Strains infecting different species appear to be a result of a limited number of capsid mutations allowing the virus to jump over the cross-species barrier. It is hard to draw parallels between the striking success of the Marion Island program and the ongoing challenges in the rabbit eradication effort in Australian. Setting aside the biological differences in the viral agents (parvoviruses are naked ssDNA viruses), the huge difference in the size of territory and the number of animals to be eliminated in both geographical locations are simply not comparable. Marion Island had only 3405 feral cats at the onset of the biocontrol program (1977) and their number was reduced to 615 (1982) in only 5 years. It took another 10 years for full eradication, until the very last cat was hunted. Looking at these two cases, one would hypothesize that earlier interventions have better chances to be successful. The chances for success are bigger when biocontrol and physical eradication can contribute to meaningful reduction of the population of the targeted invasive species.

7.2.3 Carp and CyHV-3 Biocontrol Effort

Carp fishes belong to family Cyprinidae that involves multiple species native to the big rivers of Europe and Asia. Some species have been domesticated and are farmed for food or decoration (gold fish). Today carp is growing worldwide and is considered one of the 100 worst invasive species included in the Global Invasive Species Database. Australia initiated a carp IPM plan in the early 2000s that included evaluation of the potential of cyprinid herpesvirus-3 (*CyHV-3*) as a potential biocontrol agent. *CyHV-3*, also known as koi herpesvirus (KHV) was associated with carp disease for first time in 1998 in Israel. The infection takes place through the fish skin and spreads through the body eventually damaging key organs for osmoregulation and thus resulting in a lethal outcome. Natural outbreaks of *CyHV-3* have been reported over the world with initial lethality around 70% and subsequent decline most likely due to host adaptation. The host range of the virus has been tested extensively and it appears to be restricted to cyprinids. There is no evidence for any dangers to humans handling infected fish. Due to its mode of transmission, host range, and infectivity, *CyHV-3* is considered a promising agent for carp biocontrol. Possibilities for applying *CyHV-3* and daughterless technology in parallel are being considered. The latter technology aims to genetically influence the balance between male and female offspring forcing population reduction by limiting the number of females.

Arm wrestling is often used as an analogy to visualize the challenges of biocontrol approaches applied to the eradication of invasive species and rightfully so. Multimodal strategies taking into account the dynamics of host–virus interactions and targeting population sizes from various angles are most likely to deliver meaningful results.

7.3 VIRUSES AS BIOINSECTICIDES

It has been estimated that at any given time 10^{18} insects exist on the planet representing at least 900,000 different species. The increasing global population challenges farming to increase productivity to meet food demands, whereas the health of the environment and growing resistance to insecticides are calling for broader implementation of biocontrol approaches. Insect viruses are considered a powerful biocontrol agent for the last several decades and slowly are finding their way in the landscape of modern agriculture. Just like their hosts, insect viruses are numerous and diverse. Representatives of 11 different families have been described to infect insects so far. Among them, baculoviruses are most extensively studied and thus a subject of commercialization as a potential replacement of traditional pesticides. In addition, baculoviruses are used for protein expression in insect cell culture (discussed in Chapter 4) and as a tool for transduction of mammalian cells (discussed in Chapter 9).

Baculoviruses are an attractive biocontrol agent due to their ability to infect multiple Lepidoptera insects (butterflies and moths), whose caterpillars present a major threat to economically important plants. Lepidoptera insects are widely spread and known for their fast adaptation to chemical insecticides. Baculoviruses' specificity allows for selective reduction of caterpillar populations of interest with no effect of unrelated species, thus preserving biodiversity. Most baculoviruses infect single or very limited number of related hosts. One exception is the *Autographa californica* multiple nucleopolyhydrosis virus (AcMNPV), which is known to infect approximately 30 different species belonging to several different genera of Lepidoptera. AcMNPV's host range offers the opportunity of a single agent to be used for the biocontrol of multiple insects.

Baculoviruses are enveloped dsDNA viruses with circular genome and rod-like capsids that are packaged in two distinct forms: (1) occlusion bodies (large particles containing virions embedded in a thick protein layer), mediating insect-to-insect transmission, and (2) budded virions, mediating cell-to-cell transmission. Specifics about the structure and mechanism of formation of both virion forms are discussed in Chapter 4. In nature, baculoviruses can be transmitted horizontally, within a generation with no reproduction, and vertically, between generations (Fig. 7.4). Horizontal transmission is closely associated

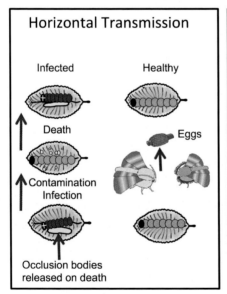
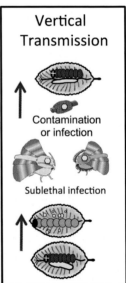

FIGURE 7.4 **Horizontal and vertical mode of transmission of baculoviruses.** Horizontal transmission occurs when larvae ingest baculovirus-contaminated food, whereas vertical transmission takes place in the context of sublethal infections persisting through the reproduction cycle. When larvae consume occlusion bodies (small *yellow dots*) and pupate before death they give rise to sublethally infected adults (marked by larger *yellow dot*), which have a chance to transmit the virus to their offspring. *Image courtesy Myers JH, Cory JS. Ecology and evolution of pathogens in natural populations of Lepidoptera.* Evol Appl 2016;9(1):231–47. https://doi.org/10.1111/eva.12328, CC-BY, https://creativecommons.org/licenses/by/4.0/.

with larval feeding. Larvae engulf occlusion bodies deposited on the surface of leaves and become infected. If the ingested dose is high, a generalized infection takes place resulting in a lethal outcome. If the ingested dose is low, larvae may become infected without developing acute disease and thus have the chance to pupate and eventually produce infected adults that can lay infected eggs. Depending on the environmental conditions, the larvae hatching from the infected eggs may develop a full-blown infection and die or develop into reproducing adults continuing the chain of vertical transmission.

By virtue of their surface proteins, occlusion and budding particles infect different cells. Ingested occlusion bodies disintegrate in the larval midgut (Fig. 7.5) and the released occlusion-derived particles attach to epithelial cells using the so-called PIF proteins (Per os infectivity factors, a group of virus proteins essential for infection of larvae, but dispensable for infection of insect cells growing in tissue culture) residing in the envelope of the occluded virus. Newly replicated virions bud through the cellular

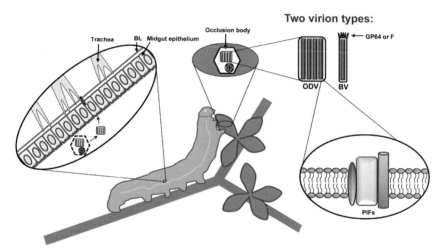

FIGURE 7.5 **Baculovirus propagation cycle on the organismal level.** Larvae get infected by ingesting leaf mass contaminated with baculovirus occlusion bodies. Occlusion bodies are protein-rich structures, which depending on the virus contain a single or multiple virions described as occlusion-derived viruses (ODVs). ODVs derive their envelope from the nuclear membrane and are released to the environment when the cells and the larvae themselves lyse. Once ingested, occlusion bodies dissolve in the alkaline environment of the larvae midgut, the ODV attach to the midgut epithelial cells using per os infectivity factor (PIF) proteins residing in their envelope. The infected midgut cells produce budding virus (BV), which carries on the cell-to-cell transmission throughout the larvae body attaching to cells with gp64 or F proteins (depending on the virus). As infection progresses, cells start producing massive amounts of BV and ODV, lyse and release occlusion bodies, which get deposited on leaves upon the larvae disintegration. *Image courtesy Clem RJ, Passarelli AL. Baculoviruses: sophisticated pathogens of insects.* PLoS Pathog *2013;9(11):e1003729.* https://doi.org/10.1371/journal.ppat.1003729, CC-BY, https://creativecommons.org/licenses/by/4.0/.

FIGURE 7.6   **Stages on baculivirus infection in *Helicoverpa armigera* (cotton bollworm) larvae.** *Helicoverpa armigera* is a moth whose larvae feed on a wide range of plants, including many agriculturally significant species. (A) wild type control; (B) dead larvae hanging in a signature wilt pose; (C) dead larvae with ruptured cuticle releasing cellular lysate rich with baculovirus occlusion bodies; (D) Baculovirus occlusion bodies at ×400 magnification. *Reprinted from Grzywacz D. Basic and applied research: baculovirus. In: Lacey, Lawrence A, editors.* Microbial control of insect and mite pests. *Academic Press; 2017. p. 27–46, with permission from Elsevier.*

membrane and infect tracheal cells using p64 or F proteins, found only in the envelopes of the budding form. As the infection progresses, the viruses spread through the entire body, get packaged in occluded form, and are released upon larval death. The progression of viral infection could be visually assessed macroscopically (Fig. 7.6). Up to 10 million occlusion bodies are released from each milligram of larval tissue, making larvae a convenient factory for manufacturing of biopesticides. In some species, the baculovirus infection results in behavior modifications that facilitate virus spreading. For example, infected larvae of the gypsy moth, *Lymantria dispar*, tend to "climb" to the top of the infested branches and die there allowing the dispersed occlusion bodies to "land" on leaves instead of the ground (Fig. 7.7). The climbing behavior is considered an evolutionary adaptation to facilitate the spread of the virus and is mediated by the viral *egt* gene coding for an enzyme involved in synthesis of a molting hormone.

Advantages of baculoviruses as bioinsecticides include: (1) safety driven from narrow host range; (2) compatibility with various modes of

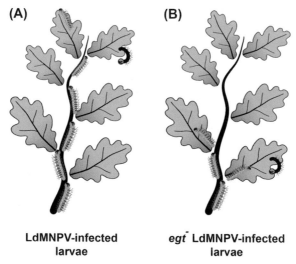

(A) LdMNPV-infected larvae

(B) *egt⁻* LdMNPV-infected larvae

FIGURE 7.7 **Illustration of the "tree-top disease" phenomenon.** *Lymantria dispar* (Gypsy moth) larvae infected with LdMNPV (*Lymantria dispar* multicapsid nuclear polyhedrosis virus) tend to climb to the top of tree branches before dying (A). The phenomenon is an example of virus-induced behavioral change, which is considered an evolutionary adaptation favoring viral spreading. Although the process mechanism is still unclear, it has been connected to the *egt* gene coding for Ecdysteroid UDP-glucosyltransferase, an enzyme involved in synthesis of molting hormones. *Egt*-mutants do not exhibit the tree-top climbing behavior (B). *Image courtesy Clem RJ, Passarelli AL. Baculoviruses: sophisticated pathogens of insects.* PLoS Pathog *2013;9(11):e1003729.* https://doi.org/10.1371/journal.ppat.1003729, CC-BY, https://creativecommons.org/licenses/by/4.0/.

manufacturing (air-drying, spray-drying or freeze-drying) and applications (spray, suspensions, powder, granules, stickers, etc.); and (3) stability at ambient temperature allowing long-term storage. Current hurdles preventing widespread use include: (1) slow rate of insect killing in comparison to chemical insecticides; (2) need for repeated application due to UV damage; and (3) high cost due to lack of infrastructure for large-scale production. Research efforts in various direction aim to solve these issues in an attempt to develop cost-effective large-scale manufacturing. In part, the slow rate of larvae lethality can be countered by higher multiplicity of infection; however, the strategy works well only in plants that can tolerate the leaf damage inflicted for several days of grazing larvae without diminishing their productivity. The approach has been found successful in trees and soybean plants whose bean yields are not drastically affected by the loss of leaf mass. Recombinant baculoviruses coding for invertebrate toxins have been engineered to speed up the action of virus-based biopesticides; however, so far none are released in nature due to public perceptions of genetically modified organisms. Virulence field studies are must before release to ensure that recombinant viruses will not be outcompeted by wild type ones naturally present in the environment. The UV-stability issue can be circumvented by repeated applications, which, unfortunately, increase cost or by including nontoxic chemicals that increase UV-stability of biopesticide formulations. At the moment, the biggest hurdle is presented from the lack of facilities and technology for large-scale production with capacity, cost, and genetic instability being the immediate problems that need solutions. These will be discussed on the example of the most extensive up-to-date application of baculovirus biopesticides developed in Brazil as a tool to combat infestation of soybean plants with velvet bean caterpillar, *Anticarsia gemmatalis*. The program utilized AgMNPV (*Anticarsia gemmatalis multiple nucleopolyhedrovirus*) and was implemented in the early 1980s on 2000 ha fields. At that time, cost-effective protocols for laboratory production of baculoviruses for large-scale biopesticide applications were not available. Soybean fields naturally infested by *A. gemmatalis* were sprayed with baculovirus and caterpillars collected 8–10 days later manually or by shaking plants on a drop cloth. Not surprisingly, these approaches generated biopesticide material with variable quality and did not have a huge capacity; however, they were essential to drive research and development efforts despite the projected high cost. Eventually, larval growth on defined media in a controlled environment was achieved and a wettable powder mixture of grinded larvae and kaolin as an inert filler was developed. By 2005, the fields treated with baculovirus biopesticide were over 2 million hectares or an 1000-fold increase in comparison to the onset of the program collectively resulting in the withdrawal of 17 million liters of chemical insecticide and 20%–30% lower cost. Detailed knowledge about the biology of the system was critical for the success of the approach. *A. gemmatalis larvae* does not disintegrate when infected and can be easily collected. Soybean plants tolerate lack of leaf mass well without significant loss of productivity, and thus the slow killing of the larvae works well for this particular case. Unfortunately, recent changes in the philosophy of soybean farming are scaling down the use of baculovirus biopesticides. Soybean seeds are being pretreated with a mixture of chemical herbicides and insecticides and then treated again with the combo several weeks after germination in an attempt to eliminate both weeds and insects with minimal number of spraying/expenses. The ecological impact of the introduced changes remains to be evaluated.

Very few non baculoviruses are understood or even researched, thus there is a huge room for field expansion and potential for development of new biocontrol agents. Recent research efforts are exploring the application of reoviruses for control of oil palm pests in South Africa; nudiviruses for control of rhinoceros beetle in Asia, and tetraviruses for control of heliothine moths. There are growing trends to optimize insect virus formulations to improve their UV stability, to increase shelf-life, and to experiment with pheromone-based and gustatory additives.

7.4 APPLICATIONS OF BACTERIOPHAGES TO CONTROL AGRICULTURAL PESTS, FOOD-BORNE DISEASES, AND FOOD SPOILAGE BACTERIA

Issues related to food production, storage, preservation, and safety are always on the forefront of technology development, safety regulations, and economic investments. Food-related diseases are a diverse group and a significant part of them are caused by bacteria susceptible to bacteriophage infections. In addition, food production and processing are complex endeavors involving many plant and animal organisms and their pests living in a constantly changing environment. Different food products are characterized by a variety of physical and chemical properties directly connected to mechanisms of their processing, consumption, preservation, and storage. As a consequence of the enormous complexity of the field, it is practically not possible to cover its depth and breadth in a fraction of a chapter, thus this section should be considered only a primer to the topic and specialized sources should be sought for a detailed description of the current state of research of phage utilization in food-related industries.

Bacteriophage applications toward control of food-borne diseases are on the interface between biocontrol and phage therapy, which is discussed in Chapter 9, dedicated to viruses as therapeutic agents. Initially, the term phage therapy was used to describe applications of phages in medicine, whereas phage biocontrol was used to describe applications elsewhere; however, the strict designation is disappearing as more authors are using the phage therapy term for nonmedical applications.

The major bacterial species causing food-borne diseases in humans targeted by ongoing phage biocontrol research are: *Salmonella, Lysteria, E. coli O157:H7, Campylobacter, Shigella,* and *Clostridium,* all causing gastrointestinal problems with a broad spectrum of severity ranging from an upset stomach to death. Outbreaks of food-borne diseases are most often associated with consumption of raw or undercooked products contaminated with the relevant pathogens. Food products of highest risk for food-borne diseases, and correspondingly of highest interest for biocontrol purposes, are meat, poultry, milk, seafood, eggs, and less frequently fruits and vegetables. The diversity of food products calls for diversity of application methods. Sprays, washing/dipping solutions, and surface immobilization are being tested as possible delivery methods. Combination applications with traditional methods for bacterial control are also in consideration. The specificity of bacteriophages makes them an excellent biocontrol tool since they can exercise their action selectively without altering the microbial community of the individual consuming bacteriophage-treated food. In addition, bacteriophage treatments do not alter the physicochemical properties of the food or its appearance. As biological entities, bacteriophages are biodegradable and thus do not pose any environmental hazard.

However, environmental factors can impact their performance significantly. Factors, such as humidity, acidity, high pressure, etc., used in conventional food preservation practices have to be taken into account when biocontrol agents are selected and tested.

Depending on the target bacteria, phage specificity can be an asset or liability. For example, biocontrol approaches targeting pathogenic *E. coli* *(E. coliO157:H7)* capitalize on highly specific viruses that target only this toxin-expressing strain and do not interfere with the growth of regular *E. coli* strains that are highly beneficial components of our microflora. On the other hand, phages with a broader specificity infecting several *Salmonella* or *Listeria* species are useful for developing broad-spectrum biocontrol products that can be universally used on various food products contaminated with slightly different bacterial strains. Development and application of bacteriophage-based biocontrol agents have to be considered in the context of relevant food processing protocols to ensure maximum efficiency. For example, it is unreasonable to apply phage biocontrol agents in environments that do not support their stability or do not take into consideration the time needed for significant action to take place. Reliable phage and bacteria monitoring methods are essential for the success of biocontrol. Even the most potent virus will not have a significant effect if it is vastly outnumbered by the bacterial host. Furthermore, bacterial species with potent restriction/modifications systems or clustered regularly interspaced short palindromic repeats (CRISPR)/Cas system are less prone to elimination and generally require higher multiplicity of infection. Biocontrol agents compatible with decontamination of equipment for food processing and storage, as well as biofilms of pathogenic bacteria are of special interest since these targets are weak spots in the food industry.

Bacteriophage applications targeting food spoilage bacterial species are also in development and are expected to contribute to increased shelf-life of food products and to reduce issues triggered by contaminating organisms in fermentation-driven processes such as dairy and beer fermentations. For example, phages are considered in biocontrol of *Pseudomonas* species, which are a major spoiling agent for milk and raw meat, *Clostridium* species associated with cheese production, *Lactobacillus* contamination of beer brewing fermentation, etc. Studies with *Pseudomonas* phages have shown that the shelf-life of packaged raw beef can be increased almost twice from 3.4 to 6.4 days, depending on the initial bacterial numbers and the viral concentration of the applied biocontrol agent. Since phages are ubiquitously spread and do not infect humans, they are considered safe and their use is not subject to regulations or reporting. Bacteriophage endolysins are another promising biocontrol tool. Similar to the bacteriophages themselves, they can vary in terms of specificity, spectrum of action, and required optimal conditions. The genes for many endolysins are cloned and their recombinant expression explored. For example, the endolysin gene of Listeria phage was engineered in *Lactoccocus lactis* as a secretable protein, which lyses contaminating Listeria without impact on the fermentation process itself.

Bacteriophage preparations are being explored as tools for biocontrol of the causative agents of common food-borne diseases at many stages of food production ranging from primary production to slaughtering, processing, and storage. Examples of commercially available products with various applications are presented in Table 7.1.

Many agriculturally significant plant diseases are caused by bacteria, and thus their causative agents are a potential target of bacteriophage-based control. The field is still making its baby steps trying to conceptualize the different hurdles to be tackled for such a strategy to become feasible, as well as to increase the inventory of isolated bacteriophages relevant to bacterial pathogens causing diseases in food-producing plants. One success story is the application of bacteriophages to control bacterial spots, also sometimes called bacterial shot holes, caused by *Xanthomonas* infection. In fact, the very first bacteriophage biocontrol product approved for use in the United States was AgriPhage produced by Intralytix Inc. to treat bacterial spots in peppers and tomatoes. Various *Xanthomonas* species and subspecies infect a broad range of agriculturally significant plants. The disease affects the entire plant causing dark spots on leaves, budding twigs, and fruit. In extreme cases, plants can be defoliated and die. The damage on the fruit also appears as dark spots altering the appearance of the fruit. Their edibility is affected only if the spots crack and suffer a

TABLE 7.1   Bacteriophage-Based Biocontrol Agents and Their Applications

Targeted Bacteria	Commercial Product/ Manufacturer	Year of Approval	Approved Application
Xanthomonas sp.	Agriphage/Intralytix Inc., USA	2005	Field biocontrol of bacterial spots in peppers and tomatoes
Listeria	LystShield/Intralytix Inc., USA	2006	Preparation of meat and poultry-based RTE foods
Listeria	Lystex P100/EBI Food Safety, Netherlands	2007	Processing aid for all foods at risk of *Listeria* contamination
E.coli, Salmonella	LystShield/OmniLytics Inc., USA	2007	Decontamination of live animals prior slaughtering
E.coli O157:H7	EcoShield/Intralytix Inc., USA	2011	Decontamination of red meat before grinding

secondary fungus infection leading to fruit spoilage. Ongoing studies in multiple plants show promising results.

The applications of bacteriophages to control plant diseases face very similar challenges as the ones already described for virus-based biocontrol processes with safety and resistance development being the most critical. An additional layer of complexity stems from the fact that the vast majority of fruits and vegetables are eaten raw thus presenting potential social acceptance controversy. Approval of phage biocontrol tools for agricultural and veterinary applications is viewed as a positive step toward revival of phage therapy practices in Western medicine.

7.5 PRACTICAL APPLICATION OF PHAGE-RESISTANT FERMENTATION MICROBES

Fermentation-driven processes are a big part of the food industry. While viruses can be a very helpful tool for the biocontrol of contaminating species in fermentation processes, they present a huge challenge as an entity attacking the fermenting microorganisms themselves and have the potential to inflict massive losses in the dairy industry. The bacteriophage threat to diary manufacturing has been recognized in the 1930s. In extreme cases, starter cultures are completely lysed and huge volumes of milk are wasted. Most frequently, phage infections slow down the fermentation process altering the quality of the final product and disrupt production schedules, thus increasing costs and lowering production capacity. For a reference, the largest dairy plants process up to 10^6 L of milk per day, which essentially grow into a huge bacterial culture of 10^{18} bacterial cells. Knowledge about the host–virus interactions between fermenting organisms and their viruses is very useful in finding practical solutions to the problem. Up-to-date research efforts have focused of isolation of bacteriophage-insensitive mutants (BIMs) from starter cultures and their molecular characterization. *Lactococcus lactis, Lactobacillus lactis,* and *Streptococcus thermophilus* are among the best-studied dairy-fermenting bacteria, commonly referred to as lactic acid bacteria (LAB), along with numerous phages that infect them. Although genetic modifications are a viable option for engineering strains with desired properties, at this moment, its significance is rather secondary as genetically modified organisms are subject of regulations and restrictions and well-established classical methods deliver results.

Milk is contaminated with phages naturally. It has been estimated that each milliliter of raw milk may contain up to 10,000 phages and that pasteurization does not kill most of them. In principle, prolonged exposure to heat is an effective tool for the inactivation of most phages; however, heat is not compatible with the dairy manufacturing processes as it impacts the quality of the final product. In addition, some of the starter culture strains harbor lysogenic phages, which when induced can lead to rapid increase of infectious phages and cause problems. Therefore it is preferable to use phage-resistant starter cultures, which do not harbor prophages to minimize potential losses.

The bacteriophages infecting LAB important for dairy manufacturing are numerous and diverse, all members of the order Caudovirales harboring dsDNA genomes. For a reference, there have been reports of 231 different phages isolated from Lactobacillus strains, most of which are siphophages or bacteriophages with long noncontractile tails. Phage resistance is best studied in the Lactoccocus system, which will be the focus of the remainder of this discussion.

Lactococcal phages are classified in ten groups, only three of which are commonly found in dairy fermentations: 936, c2, and P335 phages. The first two groups encompass only lytic viruses, whereas the third groups together mainly lysogenic ones. Raw milk is contaminated with all three groups with the c2 group being most abundant. In contrast, 936 phages are the most abundant group in fermenting milk, followed by the P335. The difference in abundance was initially attributed to different heat resistance of the groups which correlates with the isometric shape of 936 and P335 phages, and the filamentous shape of the c2 phages, respectively. Later it was realized that starter cultures contain a deletion mutant of Lactoccocus lacking the so-called *pip* gene whose product happens to be a receptor for the c2 phages. The value of this natural deletion is rather obvious giving a growth/survival advantage to the bacteria without altering its economically significant fermentation properties.

Dairy products originating from small farms that do not use commercial starter cultures but rather utilize fermentation organisms naturally found in milk are used as a source of phage-resistant strains. The rationale is that they may reflect the ecological balance between bacteria and phage in the particular environment, thus increasing the chances of phage-resistant organisms to be isolated. BIM strains are selected in the lab by employing the method of secondary culturing. Liquid cultures of sensitive bacteria are infected with a phage of interest and grown for a long time (up to 48 h). Since bacteriophages have short life cycles and experiments are planned in the context of host/virus ratio resulting into complete lysis as appreciated by the lack of culture turbidity, the approach selects for mutants that were able to escape the phage attack. Due to the low frequency of mutations, their presence cannot be easily detected in the liquid culture. Survivors, i.e., BIMs, are isolated by plating the lysate on hard media and isolating growing bacterial colonies. The BIM phenotypes are confirmed through three consecutive rounds of culturing in the presence of the corresponding phage to ensure the stability of the mutation. In most cases, BIMs contain single or multiple point mutations, which can revert back to wild type relatively easily. Occasionally, deletion BIMs as the one described above are selected and applied in practice. BIM phenotypes associated with restriction/modification and CRISPR/Cas systems have been described; however, adsorption defective mutants remain preferred since they prevent phage infection completely. While the classical method for selecting BIMs tends to be labor-intensive with a low probability of success driven by the low frequency of spontaneous mutations and the multistep nature of the process, it has proved itself invaluable as a tool to understand the biology of fermenting bacteria.

A new method for selection of absorption mutants has been recently developed. It takes advantage of the flow cytometry and fluorescently labeled antiphage antibodies and/or Hoechst 33258-labeled phages (Hoechst 33258 is a fluorescent dye that binds to DNA without sequence specificity). Bacteria and phages are mixed together for a short time to allow phage adsorption and phages are visualized directly if labeled phage was used for infection or indirectly with antiphage fluorescently labeled antibodies. Naturally arising adsorption mutants are identified by the lack of fluorescence, selected, verified, and characterized. Although the approach is expensive and requires specialized equipment and reagents, it is very fast and sensitive, opening the door for swift selection of high-performing phage-resistant strains as the need of replacement of currently used ones arises. The approach is applicable to any bacteria/phage pair pending availability of specific antibodies or successful phage labeling.

It has been appreciated that phage resistance can be established via multiple mechanisms and that starter culture strains are subject to extensive adaptation to growth in the nutrient-rich milk environment. In addition to classical methods of sanitation, the dairy industry uses media, limiting phage growth and continuously monitors the characteristics of starter cultures. The media limiting phage growth is rich in phosphates, which are meant to coordinate calcium and other divalent ions needed for phage adsorption. Known mechanisms of phage resistance fall in several major groups: inhibition of phage adsorption, preventing DNA entry, action of restriction/modification systems, CRISPR/Cas immunity, and abortive infection systems (Abi). Each of the above groups includes multiple examples of specific mechanism/players, many relying on genes encoded by plasmids. Multiple phage resistance mechanisms are often found in the same fermenting strain.

LABs are known to harbor many plasmids, which code for properties essential for fermentations as well as phage-resistance functions, known as *Abi*. Up-to-date types of 20 different Abi phenotypes have been described in *Lactococcus* differentiated with letters of the alphabet *AbiA* to *AbiU*, most of which are associated with a single gene product. *Abi* systems are not unique to *Lactococcus* and have been found in *Escherichia coli*, *Bacilus subtilis*, *Shigella dysenteriae*, among others. Abortive infections are characterized with normal phage adsorption and genome penetration followed by absent or abnormal biosynthesis resulting in the packaging and release of very few new virions, if any. The underlining mechanisms vary between different *Abi* systems and are not well understood. Known *Abi* phenotypes include delayed or absent replication, decay of phage transcripts, interference with viral recombinases and endonucleases. A proposed model for the action of the *AbiD1* system is depicted on Fig. 7.8,

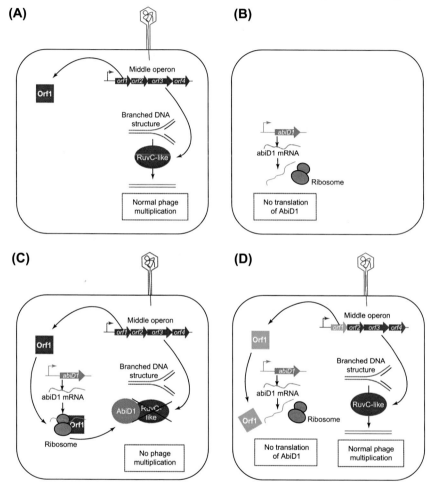

FIGURE 7.8 **Model for the action of *AbiD1*/abortive infection D1 in Lactoccocus.** Bacteriophage bIL66 codes for a RuvC-like endonuclease, which resolves branched-DNA structures during phage DNA replication and is indispensable for viral propagation. When bIL66 infects *abiD1⁻* cell the RuvC-like endonuclease is expressed and productive infection takes place (A). In absence of infection, the *abiD1* locus of *abiD1⁺* cells is not expressed (B). When bIL66 infects *abiD1⁺*, cell the bacteriophage orf1 gene product induces the expression of the *abiD1* locus resulting in inhibition of the RuvC-like endonuclease activity and arrest of phage replication (C). *AbiD1*-resistant mutants of bIL66 carry a mutation in Orf1, which is unable to induce *abiD1*, thus allowing phage replication to proceed normally (D). *Reprinted from Chopin MC, Chopin A, Bidnenko E. Phage abortive infection in lactococci: variations on a theme.* Curr Opin Microbiol *2005;8(4):473–9, with permission from Elsevier.*

showing side by side the sequence of key events in phage-sensitive and phage-resistant host, along with phage mutation overcoming the action of the *AbiD1* system. It is foreseeable that recent advances of DNA sequencing, genomics, and bioinformatics will speed up the understanding of the mechanisms of *Abi* systems and phage resistance in general.

Detailed understanding of phage resistance along with continuous identification of phage receptors will allow further advancement of the field and pave the path to rational manipulations of fermenting organisms to achieve lower incidence of phage infections and economic losses. Since the number of phages infecting dairy relevant fermenters is in the hundreds, it is unrealistic to expect that the problem will be solved permanently. The same is true for any other type of fermentation-based food production driven by bacteria. A more realistic goal is to develop algorithms for the fast and reliable phage detection and identification and a broad array of tools that can counterinteract them to be swiftly applied toward selection or generation of fermenters resistant to emerging treats.

By definition, virus-based biocontrol approaches rely on the dynamic nature of host/virus interactions in which both sides are subjects of mutations and adaptations. Detailed understanding of the interaction mechanisms and their connections with other players and factors in relevant communities and ecosystems are a prerequisite for developing effective biocontrol programs and maintaining a proactive approach to ensure a healthy environment, healthy organisms, and economic efficiency.

Further Reading

1. Buttimer C, McAuliffe O, Ross RP, Hill C, O'Mahony J, Coffey A. Bacteriophages and bacterial plant diseases. *Front Microbiol* 2017;**8**:34.
2. Di Giallonardo F, Holmes EC. Viral biocontrol: grand experiments in disease emergence and evolution. *Trends Microbiol* 2015;**23**(2):83–90.
3. Endersen L, O'Mahony J, Hill C, Ross RP, McAuliffe O, Coffey A. Phage therapy in the food industry. *Annu Rev Food Sci Technol* 2014;**5**:327–49.
4. Lacey LA, Grzywacz D, Shapiro-Ilan DI, Frutos R, Brownbridge M, Goettel MS. Insect pathogens as biological control agents: back to the future. *J Invertebr Pathol* 2015;**132**:1–41.
5. Mahony J, McDonnell B, Casey E, van Sinderen D. Phage-host interactions of cheese-making lactic acid bacteria. *Annu Rev Food Sci Technol* 2016;**7**:267–85.
6. Marco MB, Moineau S, Quiberoni A. Bacteriophages and dairy fermentations. *Bacteriophage* 2012;**2**(3):149–58.
7. McColl KA, Cooke BD, Sunarto A. Viral biocontrol of invasive vertebrates: lessons from the past applied to cyprinid herpesvirus-3 and carp (*Cyprinus carpio*) control in Australia. *Biol Control* 2014;**72**:109–17.
8. Saunders G, Cooke B, McColl K, Shine R, Peacock T. Modern approaches for the biological control of vertebrate pests: an Australian perspective. *Biol Control* 2010;**52**(3):288–95.
9. Withey S, Cartmell E, Avery LM, Stephenson T. Bacteriophages—potential for application in wastewater treatment processes. *Sci Total Environ* 2005;**339**(1–3):1–18.
10. Zaczek M, Weber-Dabrowska B, Gorski A. Phages in the global fruit and vegetable industry. *J Appl Microbiol* 2015;**118**(3):537–56.

Viruses as Tools for Vaccine Development

8.1 THE AERIAL VIEW OF THE HUMAN IMMUNE SYSTEM

The human immune system is a complex entity in our body defending us against pathogens. In essence, it employs a continuous surveillance of molecules circulating throughout the organism and uses self versus foreign triage to identify substances presenting a threat to our health. Once foreign invaders are identified, the immune system deploys an army of diverse tools to neutralize them. The human immune response is incredibly complex and swiftly adapts to meet the challenges of each day of our lives.

Figuratively, one can view the immune system as a dynamic collection of cells, tissues, organs, and processes working together to prevent disease. The immune system targets microorganisms (bacteria, viruses, protozoa), parasitic worms, toxins, allergens, and even our own cells when they display unusual characteristics. Since pathogens evolve constantly, the immune system is equipped to deal with an enormous diversity of antigens (broadly defined as any entity that triggers immune response) and to adapt quickly. Essentially, every molecule can be an antigen; however, science has figured out that proteins and carbohydrates elicit the strongest response, whereas nucleic acids and lipids are very poor antigens. The immune response is a highly coordinated and layered process that progresses from employing physical means to producing chemical and cellular tools with an increasing level of specificity. As a result, the invading pathogen is destroyed and the body acquires long-term protection and immunological memory ensuring prompt response to future encounters with the same antigen. The immune response is often described by: (1) contrasting the innate and acquired components (Fig. 8.1), (2) comparing responses to primary and secondary exposure to the same antigen, and (3) delineating cellular and humoral pathways, all of which are described briefly below after a short overview of pathogen encounter with the human body.

Harnessing the Power of Viruses
https://doi.org/10.1016/B978-0-12-810514-6.00008-8

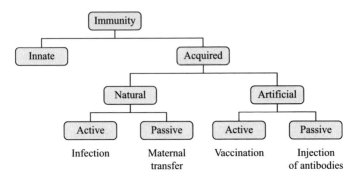

FIGURE 8.1   **Branches of the immune system.** The immune system executes its function by the coordinated action of innate and acquired immunity. Innate immunity works non-stop, whereas acquired immunity is triggered by penetration of antigens across the physical and chemical barriers of the body during infection (natural, active) or during vaccination (artificial, active). Passive acquired immunity refers to the acquisition of premade antibodies naturally via maternal transfer or artificially via therapeutic injection.

All layers and players of the immune system function simultaneously and enhance each other's action. It is not uncommon for responses to multiple antigens to take place simultaneously in the body, progressing with different pace and intensity depending on the nature of the antigen and the history of previous interactions.

8.2 OVERVIEW OF A PATHOGEN ENCOUNTER WITH THE HUMAN BODY

The immune system is very complex and versatile. It employs various organs, systems, cells, and molecules acting locally or systemically. A pathogen interacting with our body encounters a cascade of physical, chemical, and cellular barriers designed to limit its spread in the body and to "present" it to the major players of our immune system triggering a specific neutralizing response. The innate component of our immune system (Fig. 8.2) is comprised of protection tools working nonspecifically against any pathogen and do not require any level of preparation/adaptation to do their job. Generally, every pathogen experiences our skin as a physical barrier, which by the nature of its anatomy presents an extremely hard-to-penetrate protective layer. As a result, pathogens enter the human body through areas of the body not covered with skin (respiratory, gastrointestinal, and genitourinary tracts; eyes) on a daily basis or via skin cuts/breaks as they arise. Skin-free areas of the body are equipped with additional tools to nonspecifically get rid of invading pathogens such as mucus, tears, cilia, resident microflora, specific enzymes, antimicrobial peptides, etc. Any pathogens that advance

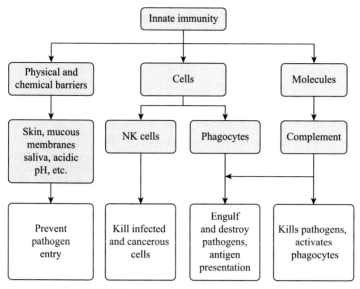

FIGURE 8.2   **Innate immunity toolset.**

further can damage or lyse cells triggering inflammation; can be identified by the complement; can be engulfed by phagocytes; or infect cells and temporarily "hide" from the immune system. Inflammation is manifested most frequently by fever, redness, and swelling, which are the net results of chemical warfare secreted by white blood cells and aimed to contain the antigen in the area of penetration/initial damage, as well as to stimulate the immune system. The complement consists of a large group of proteins that function in a chain reaction starting with pathogen identification and ending with quick microbe destruction resulting from the formation of the so-called membrane attack complex (MAC). MACs are an assembly of complement components that insert themselves in the cellular membranes of bacteria or the envelopes of viruses and create a hole/channel thus disrupting the integrity of the microbe. The complement system can be activated by carbohydrates on the microbe surface or via the Fc fragment (described below) of the antibodies reacting with the pathogen. In addition to its direct action, the complement also contributes to further stimulation of the immune system by releasing chemicals that increase vascular permeability and stimulate the proliferation of cells of the immune system. Phagocytic cells (their specific names vary depending on the organ they were initially described from) neutralize pathogens and while doing so function as antigen-presenting cells, thus bridging the action of the innate and acquired components of the immune system (Fig. 8.1). The innate component of the immune system (briefly described above) is short-lasting and nonspecific. It is usually triggered by activation of the pattern

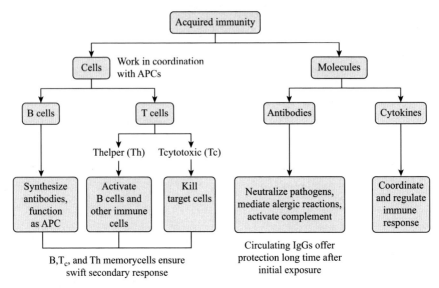

FIGURE 8.3   **Acquired immunity toolset.**

recognition receptors sensing molecules released by damaged cells or common features of microorganisms. We as individuals are born with all the tools of the innate immunity preset and working. Their encounter with antigens does not alter the respective mode/magnitude of action and does not result in any specificity. The innate immunity is poorly regulated, does not involve memory elements, and does not discriminate between self and foreign molecules.

In contrast, acquired/adaptive immunity (Fig. 8.3) is triggered by individual encounters with specific antigens. We are born with the basic tools that execute acquired immunity; however, they are in precursor form and need to be activated. Adaptive immunity is highly specific, highly regulated, discriminates between self and foreign, and results in long-term protection with strong memory element. Innate and adaptive immunity complement each other well and contribute to various aspects of pathogen neutralization and disease prevention. The innate immunity works the same way during primary (executed when an organism encounters an antigen for the first time) and secondary immune response (executed when an organism encounters an antigen for the second and subsequent times), whereas the adaptive immunity works much faster and better with each repeated exposure (Fig. 8.4). The latter is a function of the immunological memory, which allows the organism to skip the lengthy process of training B lymphocytes to produce highly specific neutralizing antibodies with high affinity. Both innate and adaptive immunity include cell-mediated and humoral components employing multiple mechanisms driven by

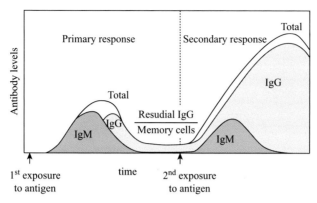

FIGURE 8.4 **Kinetics of the acquired primary and secondary immune response.** Primary immune response unfolds following the first exposure to a specific antigen. It is characterized by a delayed onset of antibody production, generation of heterogeneous IgM and IgG antibodies whose affinity increases overtime and formation of memory cells. Secondary immune response is executed following second (or consecutive) exposure to a specific antigen and results in fast and robust production of high-affinity IgG antibodies, which in most cases effectively neutralize the pathogen before disease symptoms develop.

the action of cells (cell-mediated) or soluble molecules, respectively. The action of the complement is an example of humoral innate mechanism, whereas antibody-driven processes are an example of humoral adaptive immunity. Vaccines take advantage of the immunological memory aspect of the adaptive immune response and provide the first exposure to antigen in a safe and controlled mode, thus allowing the body to undergo primary immune response and to be prepared to execute robust secondary immune response when the actual pathogen is encountered. Depending on their composition, vaccines elicit both cellular and humoral immunity, or only humoral immunity.

8.3 KEY CELLS AND MOLECULES OF THE ADAPTIVE IMMUNE SYSTEM

Several types of cells are instrumental in the execution of the adaptive immune response: antigen-presenting cells, T and B lymphocytes (Fig. 8.5). These interact with each other to accomplish three major goals: (1) antigen presentation leading to unlocking of the adaptive immunity, (2) T lymphocyte activation resulting in execution of cellular adaptive immunity, and (3) B lymphocyte activation leading to antibody production. The above processes are interconnected and highly regulated. Their selectivity and specificity is mediated by the functions of lymphocyte surface receptors, the molecules of the MHC I and II (Major Histocompatibility Complex I and II), and numerous cytokines,

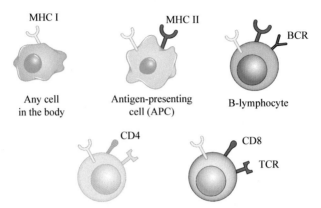

FIGURE 8.5   **Surface molecules repertoire of key cells of the acquired immune system.** The cells of the immune system employ key surface molecules to interact directly with antigens or with each other. Major Histocompatibility Complex I (MHCI) molecules are found on the surface of all cells in the body and display molecules made inside the cell. MHC II, Major Histocompatibility Complex II, molecules are found only on the surface of antigen-presenting cells (APC) and display molecules originating from the cellular environment and engulfed by phagocytosis. B and T lymphocytes display highly specific B- and T-cell receptors (BCR and TCR, respectively). T cytotoxic (Tc) and T helper (Th) lymphocytes are easily distinguishable by their specific markers CD8 and CD4, respectively, which assist the interaction of the TCR with cognate antigen presented in the context of MHC. MHCI/Tc interactions are critical for self versus foreign recognition and neutralizing infected cells, whereas MHCII/Th interactions are essential for B-cell activation.

chemicals used by the cells of the immune system for communication and regulation. In this chapter, cytokines will be referred to as a collective group without specific discussion of details of their functions. The interplay of key cells and molecules leading to antigen neutralization are discussed in the context of Influenza/Flu virus infection (Fig. 8.6).

8.3.1 Antigen-Presenting Cells

Antigen presentation is a complex process that starts with phagocytosis of pathogens. Influenza viruses that are not stopped by the chemical and physical barriers of the innate immunity reach the respiratory epithelium and carry on productive infection resulting in release of hundreds of new viruses. Some of them are released on the inner side of the respiratory epithelium and eventually engulfed by antigen-presenting cells (APCs). The formed phagosome merges with cellular lysosomes resulting in hydrolysis of the molecules building the invading pathogen. Small protein or carbohydrate pieces collectively called epitopes or antigenic determinants (roughly 8–10 amino acids or comparable length pieces of other biopolymers) are loaded in the pockets of MHC II molecules (see below) before they are exported to the cell surface.

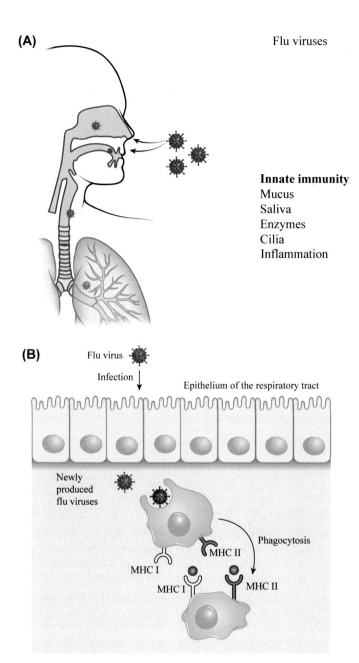

FIGURE 8.6 **Coordination of innate and acquired immune response against influenza virus.** Influenza enters the human body via the respiratory tract. If the virus is not stopped by the barriers of the innate immune system (A), it infects epithelial cells and propagates, releasing new virions across the surface layer of the respiratory tract. Lung phagocytes act as antigen-presenting cells (APCs) engulfing the virus, breaking it down, and displaying influenza antigens in MHC II. Assuming that the environment of the APC cell supports viral gene expression, viral proteins will be synthesized and displayed in MHC I (B). The activated APC cell travels to the closest lymph node, where it encounters T and B lymphocytes, effectively triggering acquired immune response.

Epitopes can be viewed as the smallest piece of molecule that can elicit an immune response. It is important to realize that each of the many biopolymer molecules present in the phagocytosed pathogen could contain multiple epitopes, thus the immune response against the pathogen is polyclonal, i.e., executed by multiple antibodies with different specificity. Once epitopes are displayed in the context (pockets) of MHC II on the surface of the antigen-presenting cell, they can be "seen" by T helper cells (Th), which can execute the next step in adaptive immune response leading to activation of B lymphocytes. MHC-II complexes are present only on the surface of APCs. In contrast, MHC-I complexes are found on the surface of all cells and they present intracellular antigen molecules. Antigens from invading pathogens presented in the context of MHC-I can be "seen" by the T cytotoxic cells (Tc) with matching receptors, which will bind to MHC I and get activated. In summary, APCs function at the interface between innate immunity and adaptive immunity. On one hand, APCs are a component of the innate immunity since they would destroy any pathogen by phagocytosis nonspecifically, but on the other hand, they could be seen as part of the adaptive response too since they are indispensable for the activation of T and B lymphocytes. In addition to displaying antigens, APCs synthesize cytokines essential for T lymphocyte activation and proliferation. APCs usually have a distinct shape with multiple protrusions allowing for efficient phagocytosis. They are found as residents of every single organ in the body as well as circulating throughout the vascular system. B lymphocytes can function as APCs themselves.

8.3.2 Lymphocytes

Lymphocytes originate from lymphoid precursors in the bone marrow and undergo maturation to naïve lymphocytes (lymphocytes that have never seen foreign antigen) in the thymus (T lymphocytes) or the bone marrow in humans/bursa in birds (B lymphocytes), from where they have gained their names. Each lymphocyte is equipped with surface receptors that could potentially recognize a specific antigen if it happens to invade the body. The specificity of the lymphocyte receptors is randomly generated in the absence of antigens. At any given moment, we have a diverse pool of naïve lymphocytes theoretically able to cover most antigens that our body could possibly encounter. The varied specificity among naïve lymphocytes is genetically predetermined by recombination events taking place in the genes corresponding to the antigen-binding sites/pockets of the lymphocyte receptors. The first encounter of a particular naïve lymphocyte with cognate antigen binding to its surface receptors functions as a selection event, i.e., the antigen-receptor binding establishes the presence of the specific antigen in the body; the lymphocyte that can neutralize the antigen is identified and selected for clonal expansion. The clonal expansion is a result of multiple cell divisions that generate a huge number of the selected lymphocyte that are ready to fulfill their function. Those lymphocytes are called effector lymphocytes, and they are actively propagated as long as the antigen is present in the body and they are needed to neutralize it. Once the antigen is eliminated, the active duty of the effector lymphocytes is no longer needed and a small percentage of them are reprogrammed into memory lymphocytes. Memory lymphocytes circulate throughout the body for years and can give rise to a new army of effector cells at any moment the respective antigen appears in the body. Each antigen can activate a miniscule number of lymphocytes from the diverse pool of naïve lymphocytes. Each activation event leaves behind limited number of memory cells. The frequency of memory cells with certain specificity is an order of magnitude larger that the frequency of the naïve lymphocytes they originated from. Memory cells have receptors with much higher affinity for the antigen compared to the one in the initial naïve lymphocyte, a phenomenon commonly referred to as affinity maturation. The affinity maturation is a function of somatic hypermutation rates and clonal expansion of the selected lymphocytes. As the amount of antigen goes down, only the lymphocytes with the highest affinity for the antigen will be stimulated to divide, thus memory cells will likely arise only from effector lymphocytes with the highest affinity.

Lymphocytes confer the ability of the immune system to discriminate between self and nonself. Precursor lymphocytes are trained to do so by extensive exposure to self-antigens early in their development. Precursors that react with self-antigens are eliminated and never become naïve lymphocytes. Occasionally, the self versus foreign discrimination fails, resulting in autoimmune disorders that are beyond the scope of this chapter. The B and T lymphocytes work in cooperation; however, T lymphocytes are mostly instrumental in cell-mediated adaptive immunity, whereas B-lymphocytes, as the sole cell type that can synthesize antibodies, drive the humoral response.

8.3.2.1 T Lymphocytes

T lymphocytes bind to antigens displayed in the context of MHC complexes using the so-called T cell receptors or TCRs (Fig. 8.7). The binding eventually results in T cell activation, clonal selection, and expansion, assuming that additional chemical signals from the APCs are received. If chemical stimulation from APCs is absent, the activation is aborted and the cells eliminated. Activated T lymphocytes synthesize various chemicals that contribute to the activation or upregulation of other players from the immune system or destruction of targeted cells.

As mentioned above, there are two major classes of T lymphocytes: T cytotoxic (Tc) and T helper (Th) cells. Tc cells selectively interact with

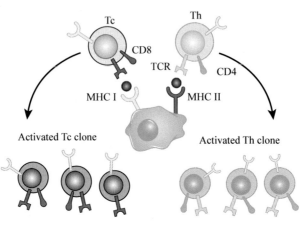

FIGURE 8.7   **Activation of T lymphocytes.** Activation of T lymphocytes is a consequence of antigen binding in the T cell receptor (TCR) and cytokine stimulation from APC. Tc lymphocytes bind only to antigens displayed in MHC I, whereas Th lymphocytes bind only to antigens displayed in MHC II. CD8 and CD4 surface markers stabilize the TCR/MHC interactions without direct contact to the antigen. The latter interactions have been omitted from the figure for simplicity.

foreign antigens presented in the context of MHC I complexes, whereas Th do so with foreign antigens presented in the context of MHC II. The ability to interact only with one type of MHC complex is a direct consequence of the fit in overall shape and chemical properties between TCRs and MHCs, as well as the presence of specific surface molecules stabilizing the interactions. Tc lymphocytes display a CD8 surface marker, which stabilizes TCR/MHC I interaction. In contrast, the CD4 marker displayed on the surface of Th lymphocytes stabilizes TCR/MHC II interactions.

8.3.2.1.1 Tc Lymphocytes

Activated Tc lymphocytes recognize and bind foreign antigens displayed in the context of MHC I. Since MHC I is present on the surface of essentially all cells in the body, an activated Tc lymphocyte can bind to any cell displaying its cognate antigen. The binding will result in release of chemical warfare, which effectively destroys the targeted cell. Tc cells are essential for neutralizing cells infected by viruses and other intracellular parasites, as well as cancer cells. Activated Th cells can stimulate activated Tc cells to proliferate in huge quantities thus helping fast elimination of cells harboring pathogens.

8.3.2.1.2 Th Lymphocytes

Unlike Tc lymphocytes, T helpers do not directly eliminate cells displaying foreign antigen; instead, they function as an indispensable and powerful assistant to the cells that can do so. Th lymphocytes get activated

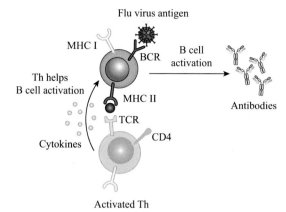

FIGURE 8.8 **Th-dependent activation of B-lymphocytes.** The binding of an antigen in the B cellular receptor (BCR) triggers its phagocytosis and antigen presentation in MHC II. Next, the displayed antigen is recognized by an activated Th lymphocyte with a complementary TCR. The formation of TCR/MHC II/antigen complex stimulates Th to release cytokines, which activate the B lymphocyte to synthesize antibodies.

as a consequence of their binding to a foreign antigen displayed in MHC II and a stimulatory signal from APC. Once activated, they start synthesizing various cytokines that assist Tc cells, many APCs (macrophages and dendritic cells) and B cells promoting their proliferation and activation. The assistance is purely chemical and takes place without physical contact. Th cells interact physically only with cognate B cells resulting in mutual activation. Depending on the nature of the synthesized cytokines, Th lymphocytes are divided into Th1 and Th2 classes, the first one modulating cellular response executed by Tc and some APC cells and the second modulating B cell–mediated humoral response. Unfortunately, the CD4 marker carried by the Th cells functions as a receptor for HIV, and Th cells are actively being destroyed as a result of the infection.

8.3.2.2 *B Lymphocytes*

The major job of the B lymphocytes is antibody synthesis. Only activated B cells can initiate antibody production. B cells can be activated directly by highly repetitive antigens or indirectly with the help of Th2 lymphocytes. The direct activation is less efficient and usually results in synthesis of low-affinity antibodies. In Th-dependent activation, the B cell initially functions as an APC displaying foreign antigen in its MHC II molecules. That antigen is bound by a Th cell, which is activated and starts synthesizing cytokines, which in turn activate the B cell (Fig. 8.8). The B cell then initiates antibody synthesis and proliferates fast, thus generating a huge population of B cells with the particular specificity. A B cell producing antibodies is commonly described as plasma cell. As B cells continue

FIGURE 8.9 **Physical neutralization of influenza antigen.** Influenza-infected cells are destroyed by activated Tc lymphocytes, which recognize the viral antigens displayed in MHC I, dock to the cell surface, and release cytotoxic molecules. The death of infected cells stops further production of new virions. Existing viruses are neutralized by antibodies, which "flag" them for phagocytosis and disrupt virus/receptor interactions thus preventing infection of new cells.

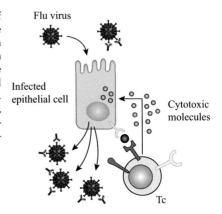

to expand, they undergo somatic mutagenesis in the genes coding for the variable regions of the synthesized antibodies and are subjects of clonal selection. Both phenomena collectively result in affinity maturation of the synthesized antibodies, i.e., molecules with increasing affinity for the antigen are made. Cytokines available in the immediate microenvironment of B cells influence what class/isotype of antibodies would be synthesized. Once the antigen is neutralized, plasma cells are no longer stimulated and they transform themselves into B memory cells ready to meet the challenge of future appearances of the antigen by synthesizing potent high-affinity antibodies.

Going back to the big picture of immune response against Influenza, the infection is contained as a result of the synergistic action of the Tc lymphocytes (cellular adaptive immunity) destroying infected cells and neutralizing antibodies (humoral adaptive immunity). Collectively, the circulating viruses are neutralized, the chain of self-propagating infection is broken and the body is on track to recover (Fig. 8.9).

8.3.3 Antibodies

Antibodies are antigen-binding immunoglobulin molecules produced by B lymphocytes. They are globular multichain and multidomain proteins with a distinct shape reminiscent of the letter "Y" (Fig. 8.10). Each antibody monomer is composed of two heavy (H) and two light chains (L), connected with disulfide bonds. Two binding sites with the same specificity are positioned at the top of the Y prongs. The binding sites are built from the N-termini of the heavy and light chains, which are highly variable between different antibodies and are designated as variable regions (V_L and V_H, respectively). The rest of the antibody molecules are fairly well conserved and designated as constant regions (C_L and C_H). The "stem" of the Y is designated as Fc fragment, whereas each prong is designated as Fab fragment. The designations

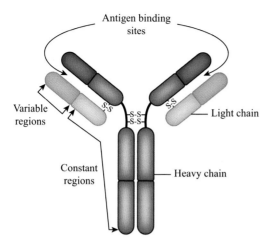

FIGURE 8.10 **Structure of immunoglobulin monomer.** Immunoglobulins are complex molecules with a sophisticated quaternary structure. Each monomer is composed of four polypeptide chains: two heavy chains (in grey), each composed of one variable and three constant regions and two light chains (in orange), each composed of one variable and one constant region. The molecules have two identical antigen-binding sites each built from one variable light and one variable heavy region.

originated from very early proteolytic studies of immunoglobulins that became established reference for the corresponding structural elements.

There are five classes/isotypes of antibodies (IgM, IgG, IgA, IgE, and IgD, where Ig is the abbreviation for immunoglobulin) found in the human body. IgD and membrane-bound monomeric IgM molecules serve as surface receptors for B cells and are regularly expressed on the surface of naïve B lymphocytes. Once a naïve B lymphocyte encounters antigen and gets activated, it starts synthesizing soluble IgM, which is pentameric and functions in the blood and the lymph during the early stages of infection/disease. IgM is a very potent activator of the complement. In the progression of infection, the concentration of various cytokines in the blood rises as more and more immune cells are synthesizing them and activated B cells undergo isotype switching and start producing a new class of antibodies. Each B cell can synthesize only one class of immunoglobulin at a time and the isotype switching is sequential: cell synthesizing IgM can switch to IgG; then potentially to IgA. IgE is associated mainly with allergic reactions and thus beyond the scope of this discussion. Isotype switching is irreversible and associated with genetic recombination. All classes of antibodies synthesized from a B cell for the duration of its life bind to the same antigen. In fact, differences in immunoglobulin isotypes are limited only to constant regions of the molecule and do not interfere with specificity.

IgG, a monomer, is the major circulating antibody in our organism, the isotype with the longest half-life and the only antibody type that can cross the placenta. IgG is a good activator of complement and up-regulator of phagocytosis. Phagocytosis of IgG-covered microorganisms is the major pathway for neutralizing pathogens.

IgA isotype is predominantly found in secretions where it exists as a dimer, whereas serum IgA is monomeric. IgA is a critical component of tears, mucus, saliva, gastric fluid, etc., where it neutralizes pathogens trying to enter the body. IgA is also found in large quantities in breast milk.

Collectively, the cells and molecules of the immune system execute complex and efficient immune response while working in a coordinated manner and mutually complementing each other. Primary immune response tends to be slow and less robust than secondary immune response. We are able to gain long-term protection against pathogens our body has already seen due to the unique ability of our immune system to support immunological memory. Vaccination takes advantage of immunological memory to trigger lasting immunity in a safe and controlled exposure to antigen.

8.4 TYPES OF VACCINES, AN OVERVIEW

Vaccines are preparations/substances aiming to provide controlled and safe exposure to infectious agent(s) with the purpose of equipping the organism with immunological tools that confer long-term protection against the specific pathogen. Vaccination, the process of applying vaccines, is one of the biggest success stories on the crossroads of science, medicine, and public health in recent history (for a timeline of vaccination and vaccine development see the Vaccine history project of the College of Physicians of Philadelphia, http://www.historyofvaccines.org/content/how-vaccines-work) that has significantly improved human health worldwide. According to the Global Vaccine Action plan of the World Health organization, today the world has in its disposal effective vaccines against 25 major diseases, some of which are caused by viruses. Thanks to coordinated worldwide efforts, small pox was eradicated, polio and measles are generally contained, and we have a global program for Flu vaccine updates. Vaccines are also a big part of veterinary practice today. The outbreak of Ebola in 2014 emphasized the need for fast and efficient approaches for development of vaccines against newly emerging highly devastating infections. The vaccine-related work of the scientific community in response to the Ebola 2014 and Zika 2016 outbreaks is good testament of the current potential of science in the field. Viruses are no longer seen simply as "the material" to produce vaccines preventing the diseases they are causing. In the 21st century, they are also becoming a tool to produce vaccines of any kind against any antigen. Other current trends in vaccine development include: (1) manufacturing vaccines targeting adults, especially elderly, rather than children only. For example, the varicella/chicken pox vaccine commonly given in early age is now also offered as an antishingles vaccine quite late in life; (2) the interest in producing multivalent vaccines is growing. They can combine vaccines against multiple pathogens in one formulation or vaccines against multiple strains of the same pathogen. For example, the anti-Flu vaccine used to include only three strains of Influenza until recently, and today a quadrivalent version is available. Similarly, the anti-HPV vaccine Gardasil initially offered protection against four strain of the virus, whereas the most updated version includes nine strains; (3) extensive efforts are focused on new routes for vaccine delivery as an alternative to traditional injections: for example, the antiflu vaccine can be taken as a nasal spray, skin patches are in clinical trials, edible vaccines are in consideration; (4) the frontiers of developing vaccines against fast-mutating viruses are being pushed forward on the example of the quest for universal antiflu vaccine and attempting new strategies for development of anti-HIV vaccine; (5) vaccine principles are being applied toward new approaches to manage chronic conditions such as high cholesterol or high blood pressure, as well as to develop therapeutic vaccines.

Vaccines can be classified by multiple criteria (Fig. 8.11) and their classification is constantly evolving as the field advances. Classical/traditional vaccines can be broadly defined as vaccines that are produced in a manner similar to well-established protocols for large-scale manufacturing. In contrast, experimental vaccines are ones in development or new to the market, usually employing somewhat nonconventional (for the time of discussion) approaches. Classical vaccines almost always induce immunity to the microorganism/microorganismal components they are made from. Heterologous vaccines are an exception of this rule and utilize the phenomenon of antigen conservation among closely related pathogens. In fact, Jenner's use of cowpox to vaccinate humans against smallpox is the best example of heterologous vaccination. Today, the antituberculosis vaccine (BCG), employed in areas of the world where tuberculosis is a frequent occurrence, uses *Mycobacterium bovis* to confer immunity against *Mycobacterium tuberculosis*.

Experimental vaccines can create immunity to a pathogen of interest while using parts of another pathogen as a vehicle. For example, recombinant vaccines may employ genetic material from the targeted pathogen delivered in VLP (virus-like particle) based on another microbe. Recently, the principles of vaccination have been applied as a therapeutic strategy to seek immunological solutions to common and devastating diseases such as cancer and cardiovascular disease. These strategies gave birth to the distinction between preventive vaccines, triggering immunity before exposure to actual pathogen causing a disease and therapeutic vaccines, triggering immune response against a disease which is already in progress. Most of the discussion in this chapter will focus on the preventive vaccines with a short overview of therapeutic vaccines toward the end.

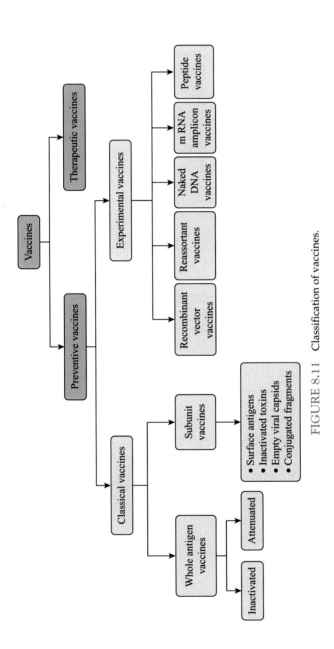

FIGURE 8.11   Classification of vaccines.

8.4.1 Classical Vaccines

Depending on their composition, classical vaccines can be viewed as whole-agent vaccines or subunit/component vaccines. Whole-agent vaccines contain a safe form of the actual pathogen causing the disease the vaccine is preventing. The infectious agent is either dead (inactivated vaccines), killed by heat, chemical treatment, etc., or alive but in weakened form (attenuated vaccines). Advantage of whole vaccines is that they mimic the infectious agent best, triggering the rise of a range of antibodies with various specificities. Dead vaccines cannot cause disease; however, they cannot induce cellular immunity either, since no productive infection takes place. Attenuated vaccines can cause full-blown infection, thus the immune response is more robust and involves both humoral and cellular components. The infection itself is very mild and does not present health concerns except in immunocompromised individuals and pregnant women. A classical approach to create an attenuated vaccine strain is to subject a wild type strain to multiple passages while monitoring virulence and immunogenicity. This strategy takes advantage of the natural ability of viruses to mutate, and depending on the mechanism of attenuation, reversions to WT virus might be of concern. Attenuated strains with deletion mutations are preferred in comparison to ones with point mutations since reversion back to WT is physically impossible. Usage of whole-agent vaccines is somewhat restricted by the ability to grow unlimited amounts of viruses in tissue culture, as well as the need for refrigeration and preservatives for long-term storage.

Subunit vaccines consist of single components of the pathogen in question, cannot cause disease, and elicit mainly humoral immune response. Subunit vaccines can be composed of inactivated toxins (toxoid vaccines against toxin-producing bacteria); surface molecules of pathogens (hemagglutinin and neuraminidase–based anti-flu vaccine); or empty viral capsids (Gardasil/anti-HPV vaccine), etc. Subunit vaccines generally require multiple booster injections to ensure strong protection. Today, most of the subunit vaccines conferring antiviral protection consist of recombinant proteins and are thus also classified as recombinant subunit vaccines. Not that long time ago such vaccines were considered experimental; however, at this moment the use of genetic engineering to produce large quantities of the protein of interest presents a rather standard practice and of course a great example of how science advancement positively impacts our life.

Conjugate vaccines could be considered as a special case of subunit vaccines, mainly applicable to bacterial pathogens. Polysaccharides from the outer surface of bacteria are attached/conjugated to inactivated protein toxins and used for vaccination. The approach solves the problem with the relatively low immunogenicity of the polysaccharides and "forces" the immune system to execute a robust humoral response against them using the toxin molecule as a tool.

8.4.2 Experimental Vaccines

Many of the classical vaccines used today were originally developed around the mid-20th century. Despite the fact that they are continuously reevaluated and refined, the magnitude of change is not even a close match to the advancement of molecular biology and biotechnology approaches relevant to vaccine design and manufacturing. While there is no obvious reason to seek replacement of effective and safe vaccines, it is only for natural new technologies to change the landscape of the field. Efforts are currently focused on creating vaccines eliciting strong cellular and humoral responses, lowering the cost of production, and increasing vaccine shelf lives, along with developing alternative delivery mechanisms. A major trend in vaccine production is moving manufacturing outside animal cells and utilizing plant, bacterial and fungal model systems. The latter are not only more cost-efficient but also eliminate the danger of pathogen cross-contamination, which is a constant concern when vaccines are produced in tissue culture, especially if mammalian cells are used.

8.4.2.1 Recombinant Vector Vaccines

Recombinant vector vaccines take advantage of molecular virology tools to produce viruses that deliver a heterologous antigen for the purpose of vaccination. Technically, recombinant vector vaccines use the same principles as virus-based gene therapy where a virus is engineered to deliver a heterologous piece of genetic information to complement existing genetic deficiency (see Chapter 9). The best example of such vaccine is the one used to vaccinate wildlife against rabies, commonly known as VR-G. The vaccine is based on a vaccinia virus vector, which was genetically modified to encode a rabies glycoprotein. The vaccine is produced by a commonly used research technology utilized in virology labs to generate mutant viruses (Fig. 8.12). In the case of the vaccine, a gene cassette encoding the rabies glycoprotein is introduced in a cell using transfection. The cell then is infected with the WT vaccinia virus and recombinant viruses are selected based on *tk* (thymidine kinase) selection. Essentially, the recombinant virus is a product of homologous recombination between the above cassette and the *tk* locus resulting in insertion of the rabies glycoprotein in the *tk* sequence. More refined versions of the vaccine replace the WT vaccinia virus with a mutant version lacking essential genes for viral propagation. Although this step complicates manufacturing, which needs to take place in a vaccinia packaging cell line providing the missing essential genes in trans, the resulting recombinant viruses are not able

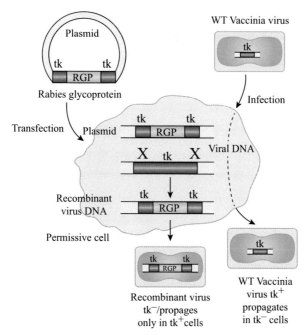

FIGURE 8.12 **Generation of recombinant Vaccinia virus vaccine against rabies.** Recombinant Vaccinia viruses expressing rabies glycoprotein (RGP) are generated by homologous recombination for the purpose of manufacturing antirabies vaccine. The RGP gene is cloned on a plasmid flanked by Vaccinia *tk* (thymidine kinase) sequences. Recombinant Vaccinia viruses arise naturally in cells transfected with the depicted plasmid and subsequently infected with wild-type (WT) Vaccinia virus. While the virus replicates, homologous recombination takes place resulting in recombinant viral genomes in which the WT *tk* gene is replaced by the *rgp* gene. Since the frequency of recombination is low, the viral progeny is a mix of recombinant and WT viruses. Recombinant viruses lack tk gene and thus can only propagate in *tk+* cells, whereas WT can propagate in *tk–* and *tk+* cells. Once selected, recombinant viruses are sequenced to confirm the replacement of *tk* with *rgp*. Large amounts of recombinant viruses are grown in *tk+* cells and formulated in the final form of the vaccine.

to replicate outside the lab, cannot revert, and are safe for widespread application. The vaccine has been used very successfully in Europe and in individual states in the United States.

The same approach was used to produce vaccines against Ebola (VSV-EBOV, licensed by Merck, Inc. and rVSV-ZEBOV, developed by researchers at National Institutes of Health, USA), which in late 2016 were in advanced stages of clinical trials. Clinical trials using adenovirus-based anti-HIV recombinant vector vaccine were discontinued due to vaccine ineffectiveness. Adenovirus-based vectors are being explored as a tool in Ebola vaccine design; bird pox viruses are considered as possible vehicles for both human and veterinary vaccines. A huge advantage of recombinant vector vaccines is their ability to simulate an infection, thus triggering strong cellular and humoral response. Unfortunately, immune response against vector antigens is also mounted, thus reducing the usability of the same vector for later applications. Nevertheless, the approach is very promising, and in theory there is no shortage of viruses that could be utilized, assuming availability of knowledge and tools necessary for their manipulations.

8.4.2.2 Reassortant Vaccines

Reassortant vaccines are manufactured by taking advantage of the natural ability of viruses with segmented genomes to reassort when more than one strain is infecting the host cell. Currently, this approach is applicable to influenza A virus, whose genome is composed of 8 ssRNA segments, and to rotaviruses, harboring a total of 11 dsRNA genomic segments. The goal of reassortment is to "assemble" a virus variant with attenuated pathogenicity that can be used for safe vaccination. Once the desired reassortant is selected, it is propagated in the context of single strain infection, thus preventing the possibility for reversion or drastic changes due to another reassortment event. Two types of reassortant rotavirus vaccines have been developed: one a reassortant of human rotaviruses, and another a reassortant of human and bovine rotaviruses. Efforts are underway to better understand the mechanism of virion packaging in influenza A, B, and C and the relevant packaging signals. It is generally thought that the three types of influenza rarely reassort due to different packing signals. Thus it is possible to utilize influenza C, which causes mild nonseasonal disease and generally does not pose a significant health threat, as a vehicle for influenza A and/or B versions of the flu surface antigens.

8.4.2.3 DNA Plasmid/Naked DNA Vaccines

Naked DNA vaccines are composed of plasmid DNA engineered to express a gene of interest, once entered into the host cell. As a nucleic acid, DNA itself has very poor immunogenic properties and is not expected to generate any immune response. Instead, the protein encoded by the plasmid would be expressed by transfected cells and secreted, thus "creating" extracellular antigen or by transfected APC cells, collectively activating both humoral and cellular immune response. As of 2016, several veterinary DNA vaccines have been licensed (for example, anti-West Nile Virus vaccine for horses, anti-Infectious Hematopoietic Necrosis Virus for salmon) and no human DNA vaccines are currently on the market. Human DNA vaccines against several viruses are currently in various stages of clinical trials; for example, influenza A, Ebola, West Nile virus, SARS. Up-to-date information about ongoing trials can be obtained by the clinical trials database run by the National Institutes of Health (ClinicalTrials.gov) Major safety concerns for DNA vaccines are: (1) the possibility for plasmid DNA integration in the cellular genome, which could potentially inactivate tumor suppressor genes and lead to cancer; (2) the possibility of triggering autoimmune diseases by generation of anti-DNA antibodies; (3) the possibility of spreading antibiotic resistance since antibiotic-based selection markers are used for plasmid production. Available data provide no evidence to support these concerns. The delivery of DNA vaccines is an area of active research experimenting with a wide range of tools, such as DNA guns, liposomes, VLPs, and polymer nanoparticles. Another area of active research is optimization of the protein expression. The biggest advantages of DNA vaccines are that they cannot cause infection of any kind, the plasmids are not transferable and cannot replicate themselves, and thus are very safe. Plasmid DNA is cheap and easy to produce, has a great shelf life, and can be transported/stored easily. The licensing of veterinary DNA vaccines, as well as the current state of the field, suggest that the first human DNA vaccines are around the corner.

8.4.2.4 mRNA Amplicon–Based Vaccines

The advancements of molecular biology and chemical synthesis fields enabled scientists to produce RNA molecules at decreasing cost and increasing length, thus making possible large-scale production of mRNAs of interest. The principle behind mRNA amplicon vaccines is essentially the same as the one behind DNA vaccine, except that the gene of interest is delivered directly as an mRNA molecule compatible with the cellular translation machinery. The advantages of RNA-amplicon vaccines are very similar to the ones of DNA vaccines; however, they erase the concerns related to genome integration and transfer of antibiotic resistance described above. Recently (2016), it was demonstrated that multiple mRNAs of interest (Influenza, Ebola, malaria) packed in a nanoparticle with dendrimeric (branched) polyamines trigger both humoral and cellular immune response, thus presenting a great candidate for vaccine. The mRNA/polyamines interaction is entirely electrostatic and the polyamines do not trigger immune response, therefore the same polyamine dendrimers can be used to pack many different mRNAs as individual vaccines or multiple mRNAs together in a multivalent vaccine. The benefits of the approach are proportional to the level of protein expression the amplicons can support, which is currently being investigated by testing various combinations of translation-related signals and regulatory elements.

8.5.2.5 Peptide Vaccines

Peptide vaccines are considered an alternative to classical vaccines that are trying to address issues of possible vaccine side effects related to vaccination with a heterogeneous multicomponent preparation. Peptide vaccines are based on in vitro–synthesized peptides of 20–30 amino acids, known to be highly immunogenic and to trigger the desired immune response. While peptide vaccines could limit significantly the chances for allergenic and/or reactogenic complications, they require carriers and adjuvants to counterbalance the low-molecular nature of oligopeptides and thus low efficiency. As of 2016, no human peptide vaccine has been approved although many are in different stages of clinical testing. The only vaccines in Phase III of clinical studies were cancer therapeutic vaccines. A veterinary peptide vaccine against canine parvovirus has been shown to be effective.

8.5 DIVA AND DISA STRATEGIES FOR DEVELOPMENT OF VETERINARY VACCINES

Veterinary vaccines and veterinary medicine face unique challenges related to disease outbreak management and transmission of vector-borne diseases. In some cases, conventional practices to contain disease outbreaks call for massive slaughtering of farm animals that test positive for antibodies against the relevant infectious agent. In such cases it is critical to be able to distinguish between seropositive animals that are sick and seropositive animals that are vaccinated. In the context of a research environment, the task is easily accomplished by PCR assays to detect nucleic acids from the infectious agent; however, at this moment it is not practical to use PCR testing in the realities of commercial animal raising. A powerful strategy to meet the challenge is to design marker vaccines that allow easy differentiation between infected and vaccinated animals. Such vaccines are commonly called DIVA (Differentiating Infected and Vaccinated Animal) or SIVA (Segregating Infected and Vaccinated Animal) vaccines. DIVA vaccines have been developed against classic swine fever, pseudorabies, foot and mouth disease, and other economically significant diseases.

DIVA vaccines are manufactured using pathogen strains lacking highly immunogenic but nonessential for propagation genes/proteins. Such strains allow for standard manufacturing process and trigger effective immune response resulting in an immunoglobulin repertoire lacking antibodies against the "removed" antigen. Serological testing probing for antibodies against the "removed" antigen offers a fast, reliable, and cost-effective way to differentiate between vaccinated animals and sick animals. Only sick animals will test positive, whereas vaccinated animals will test negative.

DISA (Disabled Infectious Single Animal) vaccines take the concept of DIVA vaccines even further. DISA vaccine is currently being developed against Blue Tong Virus (BTV), which causes hemorrhagic disease in ruminants and is transmitted by arthropods from the genus *Culicoides* (biting midges). The vaccine takes advantage from the nonessential protein NS3/NS3a, which has a rather unusual combination of properties:

(1) it is not essential for in vitro virus replication, thus allowing vaccine to be easily produced in tissue culture; (2) deletion of NS3/NS3a results in delayed virus release in mammals and highly reduced virus release in Culicoides cells, thus effectively making the virus much less virulent and much harder to transmit uncontrollably via the vector-borne route. Sheep infected with the NS3/NS3a deletion virus do not develop viremia. Ingested deletion virus does not propagate efficiently in the body of the biting midges, thus making the uncontrolled transmission even less likely. In the same time, the absence of the NS3/NS3a protein offers a convenient marker to distinguish between vaccinated and infected animals. The DIVA/DISA vaccines offer a great demonstration of how every bit of knowledge about the basic biology of processes has a great potential for practical application if placed in the right context.

8.6 NEW TECHNOLOGIES FOR VACCINE DELIVERY

Vaccine delivery is a fast evolving field focused on solving current challenges ranging from vaccine biology to vaccine economics and social aspects of vaccination. On the organismal level, vaccines are delivered almost exclusively via injection: (1) intramuscular (most injectable vaccines); (2) subcutaneous (measles vaccine); and (3) intradermal (BCG vaccine against tuberculosis). Some exceptions are the oral polio vaccine delivered as droplets, oral rabies vaccine targeting wildlife delivered as a component of attractive bait, nasal spray vaccine against influenza, DNA gun delivery.

8.6.1 Microneedle Patches

An attractive new option for delivery is the microneedle patch, shown to be effective in mice (2008) and currently in human trials (2016). The patches are composed of dissolvable microneedles delivering vaccine preparation through the skin. The skin patch is expected to help boost vaccination rates especially targeting individuals with needle phobia, the populations of rural areas, and the developing world. Assuming availability of vaccines that do not require refrigeration, it is expected that the patch can be mailed and self-applied by patients. Initial studies of public attitudes show some level of concern regarding self-application, which triggered the development of an accessory device generating a popping sound when sufficient pressure is applied to the patch.

8.6.2 Edible Vaccines

A very attractive idea for alternative vaccine production and delivery is genetically engineering plants to produce vaccines that would be delivered to the human body as part of our diet, i.e., by eating traditional fruits and vegetables. Vaccine production in plants is already a fact due to advances of molecular farming (Chapter 4). However, the available vaccines are not edible, but rather traditional injectable component vaccines manufactured in plants. The bait rabies vaccine used to vaccinate wildlife is technically an edible vaccine; however, it contains attenuated vaccinia virus strain genetically modified to display rabies surface glycoprotein, i.e., newer generation subunit vaccine delivered in an edible packaging. The latter is effective because it uses the infectivity of the vaccinia virus to penetrate the animal body and is not limited by the so-called oral tolerance of our immune system. Oral tolerance essentially allows us to eat without detrimental immunological reaction to components of our food. Once a mechanism to overcome oral tolerance is found, it is envisioned that fruits and vegetables from our diet will be used to produce the vaccines. It is envisioned that plant material will be dried and packaged in capsules for oral delivery. It is hoped that the edible vaccines will not require refrigeration and will be significantly cheaper to produce. A huge hurdle in the process is the limited number of plants that can be easily manipulated by the tools of genetic engineering. The best candidates so far are tomatoes and potatoes, which are part of the human diet worldwide and happen to be relatives of tobacco, one of the most genetically amenable systems, but unfortunately, not edible due to toxicity. Progress has been made in genetically engineering bananas. Another problem to be solved is the delivery of consistent biologically active dose. Most likely, we are decades away from mass production of edible vaccines.

8.6.3 New Technologies on the Cellular Level

A plethora of new technologies are being developed for vaccine delivery on the cellular level. Among , liposomes, niosomes, virosomes, immunostimulatory complexes (ISCOMs), polymeric and nonpolymeric particles. Use of viruses and viral capsid-based nanoparticles as delivery vehicles is discussed in Chapter 6. Collectively, these carriers of immunogenic substances are expected to be engulfed by APCs and to induce cellular and humoral responses. Each delivery vehicle has different advantages and disadvantages. Particles based on existing biological structures are biodegradable and depending on composition may, or may not, trigger immune response themselves. Synthetic particles do not induce immune reactions; however, they pose potential toxicity issues and their removal from the human body is to be studied in detail. Adjuvants (substances that increase the immunogenicity of vaccines) are another subfield of vaccine development that is actively exploring new approaches, however, outside the scope of this chapter.

8.7 USING VACCINATION PRINCIPLES TO DESIGN IMMUNOLOGICAL SOLUTIONS TO MAJOR HEALTH CHALLENGES

In-depth understanding of the inner workings of the immune system and the exploration of multiple avenues of vaccine development in recent years have advanced not only the design and manufacturing of new preventive vaccines but also initiated a variety of approaches to design immunological solutions of major health challenges such as cancer or cardiovascular diseases associated with high blood pressure and cholesterolemia.

Cancer therapeutic vaccines are aiming to elicit strong immune response against tumors taking advantage of tumor-associated antigens (TAAs). Similar to conventional preventive vaccines, scientists are exploring possibilities to inject inactivated whole cells, DNA or mRNA isolated from cancer tissues, in hopes of upregulating the immune response and changing the outcome from the continuous tug-of-war between cancer growth and organismal defense systems. Most frequently, TAAs are introduced in the body via intramuscular injection of recombinant viruses coding for TAA or displaying TAA on their surface. The resulting infection is expected to trigger a classical sequence of immune response events starting with antigen presentation and leading to production of TAA-specific antibodies and Tc cells that selectively target cancer cells. Recombinant viruses coding for a combination of TAA and cytokines are also being tested. Progress has been made in the fields of treating metastatic prostate cancer using dendritic cell–based immunotherapy (Sipuleucel-T/Provenge, approved in 2010), which is often described as a dendritic cell vaccine. Patients' peripheral blood APCs are isolated and challenged in vitro with a well-known prostate cancer antigen, as well as with growth-stimulating factors. Collectively, the manipulations result in APCs' activation and stimulation to proliferate. The generated "army" of APCs presenting the cancer-specific antigen is returned into the patient's body where they trigger cellular and humoral reactions employing the natural features of the immune system.

Therapeutic vaccines aiming to modulate different aspects of cardiovascular health employ a principle similar to the cancer therapeutic vaccines; however, instead of cells target key regulatory molecules instrumental in the control of blood pressure or cholesterol levels. For example, the angiotensin II hormone, controlling vasoconstriction, has been targeted in model animals in an attempt to control high blood pressure. The organism is challenged/"vaccinated" with the hormone peptide attached to a viral particle. The resulting immune response lowers the amount of the circulating angiotensin II, thus lowering blood pressure.

A similar approach was applied toward the PCSK9 protein, a regulator of cholesterol metabolism. Lowering the amount of PCSK9 results in lower cholesterol concentrations, and thus cardiovascular benefits. While the initial steps in the above two approaches are promising, they also raise many questions. For example, is it reasonable to use such strategies without the availability of an "OFF" switch before we know enough about the phenomenon? Periodic injection of anti-PCSK9 antibodies produced outside the body has been discussed as a valuable alternative with a straightforward "OFF" switch, although its cost and the convenience differ significantly in comparison to a vaccine. If such approaches become part of routine health care, when is a good time to introduce them: after everything else has failed or as the first intervention? Immunocontraceptive vaccines employ the same principles as therapeutic vaccines targeting cells, molecules, and processes related to reproduction. They have been explored as a tool for biocontrol and are also considered for humans. Myxoma virus–based immunocontraceptive vaccines showed great promise in laboratory trials for control of rabbit reproduction; however, field studies demonstrated that wild type (WT) strains outcompete the vaccine strain rendering the process ineffective. Without a doubt, exciting developments and likely new challenges in the field of vaccines are yet to come and virus-based tools are positioned to play an essential role in the process.

Further Reading

1. Cawood R, Hills T, Wong SL, Alamoudi AA, Beadle S, Fisher KD, Seymour LW. Recombinant viral vaccines for cancer. *Trends Mol Med* 2012;**18**(9):564–74.
2. Chahal JS, Khan OF, Cooper CL, McPartlan JS, Tsosie JK, Tilley LD, Sidik SM, Lourido S, Langer R, Bavari S, Ploegh HL, Anderson DG. Dendrimer-RNA nanoparticles generate protective immunity against lethal Ebola, H1N1 influenza, and *Toxoplasma gondii* challenges with a single dose. *Proc Natl Acad Sci USA* 2016;**113**(29):E4133–42.
3. Chen Q, Lai H. Plant-derived virus-like particles as vaccines. *Hum Vaccin Immunother* 2013;**9**(1):26–49.
4. Krammer F. The quest for a universal flu vaccine: headless HA 2.0. *Cell Host Microbe* 2015;**18**(4):395–7.
5. Liu JK. Anti-cancer vaccines – a one-hit wonder? *Yale J Biol Med* 2014;**87**(4):481–9.
6. Mennechet FJ, Tran TT, Eichholz K, van de Perre P, Kremer EJ. Ebola virus vaccine: benefit and risks of adenovirus-based vectors. *Expert Rev Vaccines* 2015;**14**(11):1471–8.
7. Norman JJ, Arya JM, McClain MA, Frew PM, Meltzer MI, Prausnitz MR. Microneedle patches: usability and acceptability for self-vaccination against influenza. *Vaccine* 2014;**32**(16):1856–62.
8. O'Hagan DT, Fox CB. New generation adjuvants–from empiricism to rational design. *Vaccine* 2015;**33**(Suppl. 2):B14–20.
9. Rubens M, Ramamoorthy V, Saxena A, Shehadeh N, Appunni S. HIV vaccine: recent advances, current roadblocks, and future directions. *J Immunol Res* 2015;**2015**:560347.
10. van der Burg SH, Arens R, Ossendorp F, van Hall T, Melief CJ. Vaccines for established cancer: overcoming the challenges posed by immune evasion. *Nat Rev Cancer* 2016;**16**(4):219–33.

Virus-Based Therapeutic Approaches

9.1 PHAGE THERAPY

Phage therapy, as the term implies, is the application of phages for therapeutic purposes. One can think about phage therapy as a biocontrol approach applying bacteriophages as means of eliminating or limiting the growth of human bacterial pathogens and thus preventing, managing, or curing the relevant diseases. In the context of human knowledge, phage therapy presents one of the very first, if not the first attempt, to use viruses for practical applications. The history of phage therapy is rather tantalizing and its future ripe to arrive pending paradigm shifts in attitudes of professional communities and general public toward virus use, genetic modifications, and clinical testing/regulations.

9.1.1 Phage Therapy: Glancing Back at History

The history of phage therapeutic applications is almost as old as the history of phage biology itself. Bacteriophages were discovered in the early 20th century independently by William Twort and Felix d'Herelle. William Twort was drawing parallels between bacteria and viruses and reasoned that nonpathogenic viruses should exist in nature, just like many nonpathogenic bacteria were known to exist. Since the number of described nonpathogenic bacteria vastly outnumbered the pathogenic ones, he reasoned that it should be straightforward to isolate such viruses if appropriate media and conditions were identified. From the perspective of current knowledge about viruses, Twort's experiments would be considered a "wild goose chase" since he tested many types of media and conditions hoping to grow viruses without a host. Ironically, the same approach applied to attempts to culture vaccinia (in search of the bacteria causing it) led to observation of bacteriophage plaques. Twort observed small glassy areas, i.e., plaques, in cultures of glycerol-preserved cowpox material that grew (were contaminated with) micrococci.

Through experimentation, he was able to establish that the material from the glassy areas contains a filterable particle that cannot be cultured on its own. The entity retained its activity for up to 6 months and could be destroyed by heat. Furthermore, the glassy areas were observed only in cultures of micrococci, to much smaller extent in *Staphylococcus aureus* cultures, and not at all in any other cultures tested (yeast, streptococci, tuberculosis bacilli). Similar glassy areas were evident when intestinal material from patients with diarrhea was cultured and instances where the culturing was obstructed due to bacterial lysis were reported. In the publication reporting the above described results, Twort made a remarkable analysis of his observations, noted the need for further inquiry to understand the phenomenon, and shared his regret for lack of funds to continue the investigation. Reading Twort's work a century later, one can easily see the budding ideas that phages or enzymes encoded by them can be used to kill pathogenic bacteria. One can also admire the research persistence and breadth of inquiry, as well as the will to share negative results.

Felix D'Herelle published his work on "invisible anti-Shiga microbe" in 1917, reporting the isolation of a filterable agent from stools of patients suffering from Shigellosis and its ability to slow down and eventually kill Shigella. D'Herelle was able to demonstrate a correlation between phage titers and the progression of disease and his work was quickly moving in a clinical direction. Initially, the idea of phage therapy was tested in the lab using Shigella infection in rabbits as a model system and in the field on the example of chicken typhoid caused by Salmonella gallinarum. These initial studies were considered very rigorous for their time, were reproduced by other scientists, and quickly paved the way to human trials. Needless to say, the 1920 practices and views on safety, clinical testing, and trajectory bench to bedside for new treatments were very different from the currently established standards. In accordance with the practices at the time, anti-Shiga phage preparations were initially ingested by D'Herelle himself, then members of his family and his coworkers. Since no negative effects were observed, the testing scheme was repeated with injection and eventually the phage preparation was given to patients with confirmed Shiga dysentery. D'Herelle was successful in treating four cases of bubonic plaque in Egypt and reducing the impact on cholera in India, thus creating momentum toward wide acceptance of phage therapy. He was also instrumental in establishing the Phage institute in Tbilisi in the former Soviet Republic of Georgia, which continues to exist today and is a major force in modern phage therapy. Phage stocks were produced by pharmaceutical companies and governments in multiple countries and applied to many bacterial infections, despite the fact that the scientific understanding of phage biology was very limited and there were no established methods for standardizing phage stocks. Not surprisingly, phage therapy was not always delivering the expected outcomes.

However, since at that time (second and most of the third decade of the 20th century) the only other option was serum therapy applicable to not that many diseases, and wars were ravaging, phages were widely used regardless of the controversies surrounding them. In the early 1930s, the Council of Pharmacy and Chemistry of the American Medical Association formally reviewed the use of phages for therapeutic purposes, and in 1934, published a report stating that their effectiveness is controversial and further research is needed to clearly establish their benefits. Around that time, the first antibiotics, sulfa drugs (1930s), were gaining ground and a few years later penicillin became available (early 1940s), thus offering powerful, easy to produce, store and administer alternative to phage stocks. The use of phage therapy gradually declined in the West, research in the area slowed down, and previously licensed phage preparations were discontinued. In Eastern Europe, phage therapy continued to be used and researched, especially in the Republic of Georgia and Poland, generating a body of clinical data whose significance is only starting to be appreciated. Meanwhile, phage biology grew into a platform of deciphering molecular biology and extensive knowledge was accumulated for key bacteriophages such as T4, T7, Lambda, and M13, among others. Bacteriophage ecology came to life and we became aware that as many as 10^{32} phages probably exist on our planet. In parallel, antibiotics usage skyrocketed, both in medicine and agriculture, and not surprisingly, antibiotic resistance gradually rose. By 2009 in the United States, more than half of the *Staphylococcus aureus* patient isolates were multidrug resistant, whereas almost 100% of the *Enterococcus faecium* isolates from critical care units were ampicillin resistant. Recent data on the impact of resistance are rather staggering. It is estimated that the cost of combating antibiotic resistance is greater than the one for combating HIV or Ebola. By 2050, approximately 300 million lives will be lost prematurely worldwide due to antibiotic-resistant bacterial infections with losses in productivity piling up to $100 trillion. Correspondingly, the World Health Organization declared antibiotic resistance "one of the biggest threats to global health, food security, and development today." (http://www.who.int/mediacentre/factsheets/antibiotic-resistance/en/), and not surprisingly, the interest in phage therapy has been on the rise since the early 2000s.

Despite the success of historically government–sponsored practices in Eastern Europe and calls for action from phage biologists, Western countries still largely disregard phage therapy as a viable clinical practice, at least in part due to the very complex landscape of regulatory, testing, and intellectual property frameworks. Many steps are being made in the right direction as reports emerge of startup companies focusing on phage therapy, animal studies, and human trials. The recent approval of bacteriophage preparations as biocontrol tools in agriculture and food industry (Chapter 7) could be seen as a segue for further developments. In 2014, the European Union issued a call for action for the development of phage therapy tools complementing traditional antibiotic treatments. In 2015, the European Medicines Agency organized a workshop on bacteriophage therapy bringing together scientists, clinicians, and politicians, among others to discuss the benefits and challenges of clinical applications of phages. The workshop concluded that bacteriophage therapy is a rather unique approach that would require a specific regulatory framework, as well as rigorous efforts to attract funding. New approaches for phage genetic modification and/or sensitizing antibiotic-resistant bacteria are being developed. There is a pressing need for goodwill and collaboration to resolve challenges faced by phage therapy. Hopefully, the scientific and medical communities as well as the society as a whole will rise to those challenges, as we have in the case of HIV and other emerging viruses, personalized medicine, and controversial technology.

9.1.2 Current Phage Therapy Practices

Active phage therapy practices are currently taking place in several Eastern European countries with work in Georgia and Poland being the most discussed. Numerous attempts have been made to analyze records of their development, testing, and applications; however, it is not clear what fraction of the applicable information is accessible due to soviet era practices, military involvement, and language barriers.

The Republic of Georgia has adopted phage therapy as part of standard health care and prophylaxis and offers to its citizens an array of over-the-counter products and specialized formulations for hospital use. Most of the phage therapy–associated activities are centered around the *Eliava Institute*, which was founded in 1918 as an Institute of Microbiology with the goal to address infectious disease issues. Its first director, George Eliava, had collaborated with Felix d'Herelle, who was instrumental in helping establish the institute as a world center for phage research and phage therapy. Up to the 1980s, the institute was manufacturing numerous phage products targeting mainly GI tract and pus-causing infections predominantly for use by the Soviet Army. After the breakdown of the Soviet Union, the institute lost the commercial scale manufacturing capabilities. It is currently producing small-volume phage cocktails for regional hospitals and collaborates with clinical staff on updating phage formulations and treatment protocols. The two major products, *Intestinophage* and *Pyophage*, were originally developed by D'Herelle in *Institut Pasteur*; however, they are subject to standard testing against relevant pathogens every 6 months and the composition of the cocktails is adjusted to meet current needs. An *Eliava Institute* spinoff pharmaceutical company, *Biochimpharm*, is licensed to manufacture and sell phage preparations in a tablet form against *Shigella* and Salmonella, including a highly specific pill against

S. typhi and has also developed a new product, *PhageBioderm*, for wound treatment. The product contains phages, enzymes, and painkillers encapsulated in a biocompatible/biodegradable polymer allowing slow release of components. Over the years, numerous studies on phage toxicity, formulation, and immune response, among others, have been performed and could be a valuable resource for establishing best practices and future policies. Phage therapy is not considered a "miracle therapy" but rather a powerful tool in a broad array of tools to combat bacterial infections with various level of complexity, in some cases (diarrhea) a prophylactic tool, as well as means for sanitation of medical environments.

Somewhat different approaches are used in Poland at the Hirszfeld Institute of Immunology and Experimental Therapy in Wroclaw. Phage therapy is considered an experimental therapy and applied only to cases, where other established treatments have failed after patient consent is obtained and approval of the corresponding regulatory entity, a bioethics commission, has been secured. The institute maintains a phage bank with more than 300 bacteriophages targeting *Escherichia*, *Klebsiella*, *Salmonella*, *Shigella*, *Enterobacter*, *Enterococcus*, *Staphylococcus*, *Pseudomonas*, *Serratia*, *Proteus*, and *Acinetobacter* bacteria. Phages are selected based on their effectiveness against the specific clinical isolate. Services are offered in a specialized institute clinic dedicated to phage therapy as well as to local/regional health facilities without any large-scale manufacturing. The outcomes from multiple clinical interventions and research inquiries are reported in more than 100 publications in several languages, as seen in the list of publications on the webpage of the institute (http://www.iitd.pan.wroc.pl/en/Phages/publications-our.html) and elsewhere. The large body of work has been reviewed recently and presents an extensive volume of observational data suggesting that phage therapy is safe and effective, as well as providing plenty of data to design meaningful clinical trials in compliance with current standards of evidence-based medicine and drug development. In addition to invaluable clinical protocols and large amount of patient data, the institute has made a sizable contribution to critical basic science areas such as phage procurement, immune response to bacteriophages, and methods for manufacturing and storage of phage preparations, among others.

9.1.3 Phage Therapy: Issues and Perspectives

Despite the urgent need for fast solutions to curb antibiotic-resistant bacteria and the many promises the phage therapy approach holds, practical advances are yet to come. Major challenges are stemming from the rather unique nature of phages as a potential medicine, absence of suitable regulatory framework, and uncertain intellectual property rights. As the first reports for clinical trials designed according to the current standards in the Western world are emerging and advances in molecular biology are applied toward engineering phages with special properties, it is becoming apparent that phage and phage products not only can successfully target bacteria and bacterial communities directly but also offer clever approaches to phage-mediated prevention of infections and even reversal of antibiotic resistance. Studies demonstrating the benefits of combining antibiotics and phage therapy approaches are starting to emerge calling for a more complex way of addressing antibiotic resistance. Four main modes of applying bacteriophages toward combat of bacterial infections, including ones with antibiotic-resistant strains, are being considered: (1) direct lysis, (2) disruption of bacterial cell walls with phage enzymes, (3) biofilm dispersal with phages coding for specialized enzymes, and (4) antibiotic sensitization (Fig. 9.1).

Direct lysis of pathogenic bacteria is a function of the host range and specificity of bacteriophages. The limited number of hosts is a great advantage

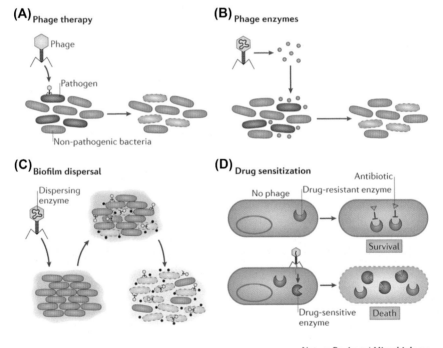

Nature Reviews | Microbiology

FIGURE 9.1 **Principle approaches of phage applications as antimicrobial agents.** (A) Phage therapy by direct lysis of phage-sensitive bacteria; (B) Bacteria elimination via phage enzymes disrupting bacterial cell walls; (C) Biofilm dispersal with phages engineered to express enzymes degrading extracellular polysaccharides; (D) Drug sensitization. *Image reprinted from Salmond GP, Fineran PC. A century of the phage: past, present and future. Nature Reviews Microbiology 2015;13(12):777–86, with permission from Nature Publishing group.*

of phages resulting in selective elimination only of the targeted pathogenic strain(s) without affecting normal bacterial flora. On the other hand, the need of production and maintaining of a broad array of different phages targeting different bacterial species or strains creates logistical challenges to the implementation of phage therapy. In that respect, bacteriophages resemble narrow-spectrum antibiotics and just like them would work most efficiently if the pathogenic bacteria are identified prior to administration of the treatment. Thus availability of fast, cheap, and reliable methods for bacterial identification in clinical and field settings will tremendously aid targeted application of bacteriophages and antibiotics. The field will benefit tremendously from fast methods to identify bacterial susceptibility to phage infection. The classical methods involve protocols that are not compatible with on-the-spot testing in the provider's office and require overnight incubations and specialized skills. New platforms for development of phage-based nanosensors have the promise to deliver both speed and sensitivity. The nanosensors are envisioned to employ phages with known specificity attached to glass slides and to be equipped to register metabolites released upon cell lysis, thus detecting bacterial strains on the time scale of a single round of phage infection (less than 30 min for most phages) without any need for culturing. While one can argue that the diversity of bacterial pathogens and specificity of phages in combination present a tough case for broad implementation of phage therapy, it is important to realize that there is limited number of bacterial strains that are largely responsible for most of the antibiotic-resistant infections. For example, 70% of MRSA (Methicillin-resistant *Staphylococcus aureus*) infections are due to only five strains of *S. aureus*, thus the task is manageable as long as good diagnostic and phage tools are available. The magnitude of the task would be reduced even further if the differences between strains are well known and variants of the same phage can be engineered to infect each strain, thus reducing the need for developing multiple modification, manufacturing, storage, etc. protocols. One can reason that targeting persistent nosocomial infections with phages or combination of phages and antibiotics should be a priority in the efforts to curb the spread of antibiotic resistance.

Direct lysis is associated with potential side effects driven by release of highly toxic substances such as endotoxins from gram-negative bacteria, a problem also associated with many antibiotics. Different strategies have been proposed to minimize the possibility of such side effects. Phages have been engineered to not lyse the cell, as well as to deliver bacteriotoxic agents such as restriction enzymes or holins. Initial experiments demonstrate that nonlytic phages deliver a similar therapeutic effect with lower level of inflammation in comparison to lytic phages. Endotoxin contamination is a general concern for manufacturing phage preparations; however, so far it has been shown that existing methods for endotoxin removal work efficiently for preparation of phage stocks.

The therapeutic effect of phages is highly dependent on their persistence in the affected areas. So far, many positive results have been reported in localized applications to wounds, the GI tract, and otitis infections; however, very limited experimentation has been done for systemic infections requiring bloodstream delivery. Viruses injected into the bloodstream are quickly cleared from the reticuloendothelial system. It has been shown that serial passaging in vivo can be used to select mutants that persist longer in the body. A bacteriophage lambda mutant named lambdaArgo carries a single amino acid mutation in its capsid protein that allows 1000-fold increase in ability to persist 24h postintraperitoneal injection. Phage surface modification with PEG increases persistence and lowers immune response. Phage display and in vivo biopanning approaches are great tools for selection of persisting phages. The issue is being actively researched in the context of using viral capsids as nanocarriers for drug delivery and any general principles elucidated in such studies can be applied to phage therapy. Phage preparations have been formulated and delivered safely in multiple forms, ranging from phage lysates for oral application, to aerosols and topical creams. Most phages tend to be stable for at least a year if prepared and stored properly.

One of the biggest concerns for the direct lysis approach is the development of antiphage resistance. Bacteria propagate very fast and mutate fairly frequently when selective pressure is applied in vitro. Experiments show that in the context of a research lab, resistance can arise within several hours or several days, when bacteria and phages are cultured together. Bacteria become phage-resistant via multiple mechanisms, such as mutating receptors that do not allow phage adsorption, restriction-modification systems, Abi (abortive infection) systems, CRISPR/Cas, toxin–antitoxin systems, etc. Some of the above resistance systems are plasmid-encoded and can be horizontally transferred between bacteria via conjugation. Utilization of cocktails of multiple phages, combinations of phages and antibiotics, or filamentous phages that inhibit conjugation have been proposed as possible solutions. Engineering modified phages targeting resistance mechanisms is another approach, whose utility has the potential to grow as those mechanisms are better understood. Since both bacteria and bacteriophages propagate very fast and continuously adapt to each other, resistance to phages is not completely avoidable; however, studies show that its frequency in vivo is much lower than the experimentally observed in vitro, thus possibly making it avoidable if multifactor approaches and close monitoring are in place. Current practices in both Poland and Georgia involve critical evaluation of used phages and phage cocktails every 6 months, followed by adjustments of strain composition and isolation of new phages on an as-needed basis. Resistance by its nature is a product of adaptation and evolution of living matter and not unique to phages or antibiotics. Examples from different fields demonstrate that no single measure for curbing growth of unwanted pest is foolproof in the long run and that the best solutions usually come from multimodal approaches that take into account ecological and evolutionary relationships between the involved entities. Similar resistance issues are inherent to HIV treatment, where early onset of treatment, application of drug cocktails, and prophylaxis are considered reasonably effective tools to combat resistance. The dairy industry uses routine monitoring for phage sensitivity and redefines starter cultures periodically to curb losses. Biocontrol approaches using viruses apply them in combination with other synergistic tools and periodically introduce new strains. In parallel, research efforts are deciphering the nature of interactions, accumulating knowledge regarding the mechanisms of ongoing processes, and eventually developing future tools. Very few pathogens and pests have been eradicated in human history. Most of the success of medicine and agriculture dealing with existing threats comes from periodic discovery of new approaches, which can deliver cost-effective results within an acceptable time frame. In that respect, phage therapy should be considered a new (old) tool to combat bacterial diseases in the face of antibiotic resistance getting out of control, as opposed to antibiotic replacement that is expected to deliver results with no adjustments over time.

An alternative strategy to direct lysis is the use of phage enzymes that interfere with the structural integrity of bacterial cells. The most researched group of enzymes is endolysins, which depolymerize the proteoglycan of the bacterial cell wall during the release of the new viral progeny at the end of the lytic viral infection. Endolysins execute their function by employing separate proteoglycan-binding and hydrolyzing domains, which are fairly specific. Surprisingly, some endolysins are effective against a broad range of bacteria. For example, the PlySs2 endolysin isolated from *Streptococcus suis* is effective against several other *Streptococcus* species, MRSA, *Staphylococcus epidermidis,* and several *Listeria* species. Mice infected with MRSA and *Streptococcus pyogenes* have been treated successfully with PlySs2 endolysin. Genetically engineered endolysins, combining domains with multiple specificities, have been shown to be effective against multiple species. The LysAB2 endolysin, isolated from *Acinetobacter baumannii,* has a broad specificity and can neutralize multiple gram-positive and gram-negative bacteria. While enzymes with broad specificity are cost effective from a research and development standpoint, they might impact nonpathogenic bacteria too, thus altering the normal microflora in the way broad-spectrum antibiotics do.

Many bacteria are living in complex communities collectively referred to as biofilms. Unfortunately, biofilms easily form on commonly used medical equipment, especially pieces that include tubing and slow-moving liquids. *Pseudomonas aeruginosa* and similar species are notoriously resistant to antibiotics and very hard to eliminate as a threat in long-term care facilities and operating rooms. In addition to the conventional mechanisms of antibiotic resistance, biofilms have complex extracellular structure that presents a physical barrier to antibiotics, thus its disruption can contribute to higher antibiotic effectiveness. Bacteriophages modified to code for enzymes degrading the extracellular polysaccharides function as a two-prong weapon that kills the biofilm bacteria by lysis and disperses the biofilm structure, thus altering the ecological microenvironment and decreasing the colonization capabilities of the bacteria comprising the biofilm. In addition to biofilm-dispersing enzymes, bacteriophages have been engineered to deliver antibiotics and photosensitizing drugs that offer high-dose targeted treatment and thus higher effectiveness. Photosensitizing drugs are activated by irradiation of bacteria with light with particular wavelength.

A unique approach to combating antibiotic resistance in bacteria is drug sensitizing, which essentially can be considered as researcher-guided horizontal gene transfer in reverse. Normally, genes associates with antibiotic resistance can be transferred between bacteria via temperate phages, conjugation, or less frequently by transformation of freely floating DNA pieces released upon bacterial death. When wild-type genes are reintroduced in bacteria to replace mutated versions conferring resistance, sensitivity to antibiotics is restored. For example, resistance to streptomycin and nalidixic acid was reversed by introducing the wild-type rpsL and gyrA genes, respectively, in E. coli using lambda bacteriophage. RpsL codes for a ribosomal protein, which frequently mutates when bacteria are treated with streptomycin, whereas gyrase mutations confer resistance to quinolones, including nalidixic acid. In a separate study, Lambda bacteriophage was instrumental in the replacement of a mutant efflux pump conferring resistance to multiple antibiotics with the wild-type sensitive protein. Several approaches employing the CRISPR/Cas system have been developed to sensitize bacteria. For example, phagemids based on bacteriophage FNM1 were engineered to deliver sequences targeting genes associated with antibiotic resistance in S. aureus. Subsequent phagemid infection effectively eliminated plasmids conferring resistance, thus sensitizing bacteria. In an alternative strategy, CRISPR/Cas components targeting genes associated with antibiotic resistance and infection with bacteriophage T7, a lytic phage, were delivered in E. coli using bacteriophage Lambda. By design, the approach genetically linked bacterial sensitization to resistance to bacteriophage T7, thus resulting in a bacterial population sensitive to antibiotic and resistant to lytic phage. The resulting bacterial population was treated with a combination of T7 and antibiotic, effectively killing all cells. The strategy effectively alters the proportion of antibiotic-sensitive bacteria in the local environment, thus balancing out the selective pressure resulting from antibiotic application. It is envisioned that the temperate/lytic phage combination can be used as a disinfectant in health care facilities to actively rebalance bacterial communities and eventually eliminate them. The approach does not include direct application of phages to patients, which may facilitate its adoption. Fig. 9.2 depicts three different approaches to attack antibiotic-resistant bacteria and predicted outcomes. Any of the three approaches could target one bacterial species at a time, thus phage cocktails will be needed to neutralize antibiotic-resistant bacteria in real-life conditions. It is important to realize that currently science does not understand well the natural interactions between animals, bacteria, and bacteriophages. Most of the studies attempting to identify phages present on the surface of the skin and mucosal tissues, are very

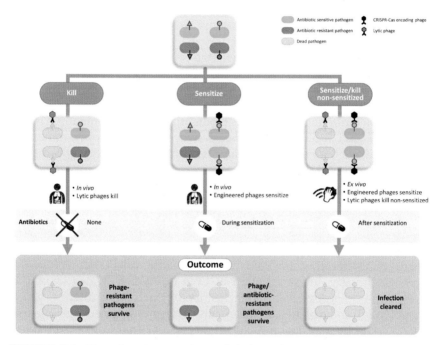

FIGURE 9.2 **Phage-based approaches to fight bacteria involving drug sensitization.** Pathogen population (top) consists of antibiotic-sensitive (green) or antibiotic-resistant (orange) bacteria, prone to infection with specific phages (*circle* and *triangle* receptors). Conventional phage therapy (left) targets bacteriophage sensitive pathogens, regardless of antibiotic sensitivity. CRISPR-Cas therapy using engineered phages (middle) sensitizes pathogens to antibiotics. When combined with antibiotics, the treatment efficiently eliminates all bacteria with the exception of phage- and antibiotic-resistant ones. Alternative approach uses CRISPR-Cas sensitization to antibiotics combined with lytic phages to kill nonsensitized pathogens ex vivo (right). This results in sensitizing the bacteriophage-sensitive population and killing those that are not sensitized. This treatment constitutes a counterselective pressure to antibiotic resistance on hospital surfaces and hands of medical personnel and does not involve direct application of phage preparations to the patient. *Image and figure legend reprinted from Goren M, Yosef I, Qimron U. Sensitizing pathogens to antibiotics using the CRISPR-Cas system.* Drug Resistance Updates 2017;**30**:1–6, *with permission from Elesevier.*

fragmented utilizing only single bacterial strains to screen for phages and not surprisingly reporting that no phages or limited number of phages are found. Systematic broad-spectrum studies have a better chance to identify bacteriophages residing on bodily surfaces, which eventually can be used for the purpose of phage therapy or engineered to deliver therapeutic agents and/or sensitize bacterial populations. There is a pressing need for a detailed understanding of how different bacteriophages interact with the immune system and how the specific interactions influence their potential as therapeutic agents. Current data point to reasonable tolerance of the immune system for bacteriophages. Ironically, not that long time ago, the dairy industry entertained the idea to address phage sensitivity of the fermenting starter cultures by immunizing the cows with phage components in hopes that high amounts of antibodies will be secreted in milk and will neutralize contaminating phages. The approach was found unfeasible, however detailed knowledge regarding phage proteins triggering immune response was accumulated, demonstrating that very few were produced. Although such studies are rather impractical in humans, chances are that there will not be drastic differences in phage immunogenicity. Bacteriophages have great potential to help combating bacterial diseases and antibiotic resistance; however, time and open minds are needed to drive the necessary clinical testing and regulatory approval to bring them to the clinics in the Western world.

9.2 ONCOLYTIC VIRUSES

Oncolytic viruses (OVs) are viruses that selectively infect and destroy cancer cells without excessive damage to healthy cells (Fig. 9.3). Similar to phage therapy, the application of OVs in medicine (also known as oncolytic virotherapy) has a long history of fluctuating interest, growing potential and recent advancements. The first data suggesting that viruses could be a potential tool to fight cancer came from observations that cancer patients went into a temporary remission when experiencing illness of viral origin or were undergoing vaccination with virus-based vaccines. These observations correlate with reports dating back to the 19th century (before viruses were discovered) that infectious bodily fluids from humans and animals with certain diseases (mumps, hepatitis) were given to cancer patients in attempts to inflict therapeutic benefits. Such practices had very limited success and quickly were discontinued. Conclusive evidence according to the modern standards was obtained in the mid-20th century when it was shown that mouse sarcoma tumors can be selectively destroyed by Russian Far East encephalitis virus. As one can imagine, the mice died from encephalitis clearly pinpointing the need to separate oncolysis and pathogenicity. With the advancement of molecular

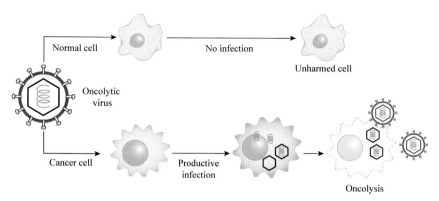

FIGURE 9.3 **Conceptual principle of oncolytic viruses.** Oncolytic viruses selectively infect and propagate in cancer cells inflicting no harm to normal cells. Selectivity is driven from the complementary nature of the specific properties of the virus and the cellular environment allowing for productive cell entry (based on viral ligand/cellular receptor interaction) and effective viral replication evading the cellular antiviral defenses.

biology, detailed characterization of pathogenic viruses became possible, enabling understanding of the natural oncolytic properties of viruses and rational modifications for the purpose of designing oncolytic viruses. The first formally approved OVs were reported in Latvia (2004, natural ECHO-7 strain of human enterovirus) and China (2005, genetically engineered adenovirus H101) for the treatment of melanoma and head and neck cancers, respectively. The first OV approved in the United States and Western Europe was reported in 2015 (T-vec, Amgen; a modified Herpes Simplex Virus Type 1) for the treatment of melanoma patients with inoperable tumors. In 2016, there were approximately 40 clinical trials recruiting patients with various cancers for protocols employing herpesvirus, adenovirus, vaccinia virus, measles virus, and poliovirus, among others.

9.2.1 Mechanism of Action

Oncolytic viruses are a unique and multidimensional class of therapeutic agents with rather diverse mechanisms of action. They combat cancer directly, selectively killing malignant cells and/or inducing anticancer immune response, or indirectly lysing endothelial cells of the tumor vasculature and thus depriving tumors from oxygen and nutrients (Fig. 9.4).

Some viruses are naturally oncolytic and preferentially infect cancer cells with specific characteristics such as: ras mutations (reovirus); interferon response–related mutation (vesicular stomatitis virus [VSV], Newcastle disease virus); presence of specific cell surface receptors (poliovirus/CD155; measles virus/CD46 or CD150); malignancy-driven cellular changes (extracellular matrix alterations promoting herpes simplex virus type 1 [HSV-1] infection). Most of the OVs are genetically

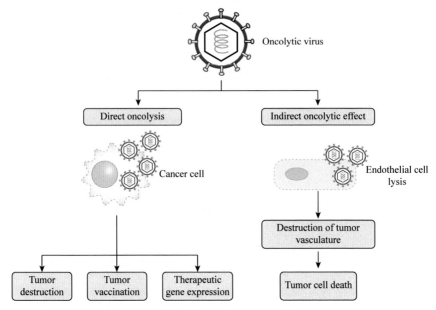

FIGURE 9.4 **Mechanism of action of oncolytic viruses.** Oncolytic viruses kill cancer cells by direct oncolysis (left) in which cancer cells disintegrate as a consequence of productive virus infection, releasing tumor-associated antigens that trigger antitumor immunity. Genetically modified oncolytic viruses code for genes that can enhance direct oncolysis most frequently by stimulating the engagement of the immune system. Viruses selectively attacking tumor vasculature kill cancer cell indirectly by depriving them from oxygen and nutrients.

engineered and thus can be considered gene therapy agents targeting cancer. Several rational design approaches to engineering have been employed such as alteration of viral surface molecules to ensure OV attachment to cancer cells, deletion of genes essential for propagation in normal cells, placing essential viral genes under the control of tumor-associated transcription factors, expression of molecules stimulating immune response. Another strategy of identifying OVs is conventional selection. For example, a pool of recombinant adenoviruses was generated by coculturing several types of adenoviruses under conditions promoting recombination. The pool was then tested for the ability to propagate and lyse different types of cancer cells, and the most potent viruses were selected. A second round of selection was performed in normal cells to exclude viruses replicating in healthy cells. An OV in clinical testing known as ColoAd1/Enadenotucirev/EnAd was selected for its ability to kill colorectal cancer cells with high potency and specificity. Molecular analysis demonstrated that the recombinant virus is based on Ad11p and Ad3 adenovirus types and harbors several deletions; however, the precise mechanism of its action remains to be understood.

Development of successful OV-based therapies is a rather challenging task as OVs are expected to fulfill multiple requirements. Theoretically, an ideal OV should: (1) preferentially target cancer cells; (2) be nonpathogenic to humans or cause mild and self-limiting disease; (3) exclusively or at least preferentially replicate in malignant cells and efficiently kill them; (4) be able to engage the immune system and trigger/enhance antitumor immunity; (5) be genetically stable in a tumor environment with minimal chances for recombination-driven alterations; (6) have effective antiviral drugs available to control infection if needed. Additional requirements are posed from the realities of virus manufacturing, storage, and safety regulations in clinical settings. Understanding the mechanism of action of existing OV and complementing their action with targeted engineering and/or supplementary approaches could lead to practical measures to meet most of the above requirements.

Current studies of OVs' mechanism of action point to the conclusion that their therapeutic effect is due to a combination of selective cancer cell killing and development of antitumor immunity. Tumor selectivity is a function of several factors, such as specific viral entry in target cancer cells, rate of metabolic activity capable of supporting productive viral infection, and deficiencies in antiviral defenses. Very few wild-type viruses are considered good candidates for virotherapy. A common feature of the ones being explored is their limited human pathogenicity, which is a natural characteristic of some human viruses such as coxsackieviruses and reoviruses. Other viruses, although pathogenic in nonhuman hosts, do not cause notable diseases when infecting humans (avian Newcastle disease virus; rat Hi1 parvovirus, vesicular stomatitis virus infecting cows, pigs, horses, insects). Taxonomically, OVs in consideration are very diverse and range from viruses with very small genomes (H1 parvovirus, 5 kB ssDNA genome) to large and complex viruses, such as HSV-1, a dsDNA virus with a genome of 150 kB. Another approach to the pathogenicity aspect of OVs is using well-established vaccine strains that have been in clinical practice for years; for example, Sabin Poliovirus or Edmonston measles strain, or using viruses that are heavily researched as viral vectors for gene therapy; for example, adenoviruses and herpesviruses.

OVs' targeted entry is a function of the nature of the viral receptor and the characteristics of the particular type of cancer cell. Since cancer cells are genetically unstable and heterogeneous, information regarding their characteristics could be crucial for meaningful selection of an OV approach. It is foreseeable that in the near future, surface receptor screening will be part of the protocol for OV treatment, similar to the currently established practices for Herceptin treatment of HER2-positive breast cancers. The poliovirus enters the cell through the CD155 molecule expressed on the surface of glioblastoma, colorectal carcinoma, melanoma, and some breast cancers, among others. CD155 is a cell surface adhesion molecule functioning in tumor cell migration, invasion, and metastasis, and not surprisingly, is also designated as a common tumor-associated antigen. In contrast, CAR (coxsackievirus and adenovirus receptor) is expressed variably in tumor cells and normally found in several tissue types. In addition, its expression in endothelial cells can be somewhat modulated with drugs (lovastatin, a statin-based cholesterol-lowering drug). Correspondingly, altering the receptor specificity of adenoviruses is being considered as an approach to increase the utility of adenoviruses as OVs. A rather interesting approach of redesigning surface molecules was employed in the measles virus. The virus uses two glycoproteins, hemagglutinin (H) and fusion protein (F), to execute a productive viral entry. Researchers engineered the H protein to prevent its binding to its cognate receptor, CD46, and fused single-chain antibodies or growth factors to its C-terminus. The fused proteins bind to surface molecules commonly displayed on cancer cells, directly guiding the recombinant virus to attach to them. The F protein was also altered by inserting a linker between its F1 and F2 domains abolishing its activity. The linker harbors a metalloprotease cleavage sequence, and when cleaved, allows for F protein maturation and subsequent activity. Since metalloproteases are known to be abundant in tumors, the F protein function will be preferentially restored, allowing efficient entry of the virus and subsequent oncolysis.

Aside from receptor-mediated entry via receptors selectively expressed on the surface of cancer cells, effective discrimination between normal and cancer cells has been achieved with the deletion of viral genes essential for propagation in normal cells but dispensable in cancer cells, due to their metabolic and genetic characteristics. For example, cancer cell selectivity of adenoviruses can be engineered by removing their E1a and/or E1b genes, which are known to interact with p53 and Rb cell cycle/apoptosis regulator proteins. When infecting normal cells, adenoviruses have to disable cellular apoptosis pathways to propagate effectively. In most cancer cells, the p53 and Rb proteins are already mutated and not working properly, thus the functions of the E1a and E1b proteins are no longer needed. Adenovirus lacking E1a or E1b effectively propagate in cancer cells leading to cell lysis and opportunities to infect the neighboring cancer cells. In contrast, normal cells do not support their propagation. VSV-based oncolytic viruses exploit the interferon antiviral defense system to discriminate between normal and malignant cells. Interferons are signaling protein molecules that drive virus-induced protein shutoff in infected cells and the tissues surrounding them (among other functions), thus preventing viral propagation. Many viruses have evolved to evade the interferon defense by developing specific bypassing strategies. For example, the VSV virus overrides the interferon defenses by the function of its matrix protein, which blocks mRNA transport, thus preventing interferon mRNA export from the nucleus and interferon synthesis and secretion, signaling viral infection. VSV with matrix protein deletion is effectively neutralized by normal cells and propagates effectively in cancer cells with disabled interferon signaling, thus offering an mRNA transport/translation-dependent mechanism of cancer cell selectivity. Transcription-based mechanisms of selectivity include the utilization of tissue-specific promoters controlling the expression of key genes for viral propagation. Tumor selectivity could be enhanced by intratumoral injections of OVs; however, the ability of OVs to survive systemic application and deliver effective results is a huge asset in terms of eliminating metastases.

9.2.2 Armed Oncolytic Viruses

Armed oncolytic *viruses* are genetically engineered OVs coding for a therapeutic gene(s), whose expression is fully dependent on the replication of the *virus*. One can think about armed OVs as enhanced OVs that have been equipped with additional means to combat cancer by employing the conceptual principle of gene therapy. A broad array of genes is under consideration as therapeutic genes arming OVs. Tumor suppressor genes (p53, Rb, p21, p27 involved in cell cycle regulation) are intended to counteract loss of tumor suppressor genes and thus induce apoptosis of cancer cells, and block proliferation and metastasis. Cytokines and other immunostimulatory genes are aiming to enhance the systemic immune response and antitumor immunity. Prodrug-activating genes/suicide genes code for enzymes that convert prodrugs into active drugs killing the cell that harbors them. For example, an oncolytic virus armed to deliver cytosine deaminase, an enzyme catalyzing the deamination of cytosine to uracil, can convert 5-fluorocytosine into 5-fluorouracil, which is a potent chemotherapeutic and radiosensitizing drug. It is hoped that armed oncolytic viruses alone or in combination with other therapeutic approaches will offer new options for cancer management.

9.2.3 Oncolytic Viruses and the Immune Response

The human immune system is a very dynamic and highly adaptable entity, designed to execute a robust inflammatory response to viruses invading the human body. Although inflammation is a potent defense tool, if excessive and prolonged, it can also severely damage normal tissues. Under normal circumstances, our body quickly extinguishes inflammation as other components of the immune response are upregulated and an army of neutralizing antibodies is manufactured, executing long-term protection. Oncolytic viruses are no exception, thus the actions of the immune system directly influence the OV effectiveness. In addition, tumors present a unique immunological environment, where many of the normal functions of the immune system are suppressed or altered, collectively allowing malignancies to persist despite the fact that the body is in principle capable of eliminating them. In most cases, the process is driven by overexpression of cytokines and chemokines altering the behavior of immune cells. For example, tumors often express IL-10 (interleukin 10) and TGF beta (transforming growth factor beta), which are potent inhibitors of antitumor immune response. OVs are often engineered to express interleukins stimulating antitumor response, in hopes of influencing the tumor microenvironment and alleviate the suppression.

While executing their cytotoxic effect, OVs have the potential to function as antitumor vaccines. Cell lysis releases tumor-associated antigens along with cellular components triggering immune response by activating antigen-presenting cells. The potency of antitumor immunity has been found to be variable between tumors and employed OVs. To capitalize on the idea, some groups have employed a prime/boost scheme with heterologous vectors, one of which is an oncolytic virus, thus combining the benefits of the principles of vaccination and oncolysis. In the "prime" round, animal models are vaccinated with a nononcolytic virus coding for dominant TAA relevant to the targeted cancerous growth. A few weeks later the "boost" round is executed by systemic injection with oncolytic virus engineered to express the same TAA, i.e., a booster. It has been found that oncolysis in the context of the prime/boost approach triggers antitumor immunity better than oncolysis alone. Whereas, engagement of the immune system is highly desirable when oncolytic viruses are employed, the immune system presents multiple obstacles to persistence of OVs in the body, which could diminish their potency (Fig. 9.5). Preexisting neutralizing antibodies could easily neutralize most of the oncolytic vectors introduced in the body. This possibility is circumvented by selection of viruses with low immunogenicity and low chances for preexisting immunity. Measures such as modifying virus surface by

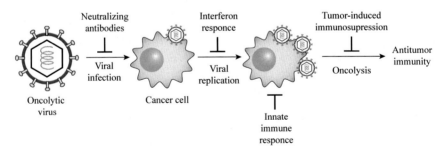

FIGURE 9.5 **Immunological barriers to oncolytic viruses.** Oncolytic viruses face the same immune system challenges as any virus invading the human body and depending on the specific circumstances can be neutralized easily or persist in the body for a while exerting their oncolytic effect. The design of the oncolytic viruses and the treatment protocol can significantly influence the effectiveness of oncolytic virus therapy.

attaching inert polymers (PEG) or by genetic engineering could contribute to avoidance of existing neutralizing antibodies. Another strategy is the use of cellular vectors for example, mesenchymal stem cells (MSC). MSC are known to have natural tropism for malignant cells and to associate with well-established tumors, as well as metastases. MSC cultured in vitro and infected with oncolytic virus are delivered in the bloodstream, reach the cancer cells, and release the oncolytic cargo in situ, thus delivering high dose of newly produced viruses in the immediate surroundings of the target cells ensuring direct infection without prolonged circulation in the blood stream, thus effectively avoiding neutralizing antibodies. Interferon response is another immunological tool that could diminish the effectiveness of oncolytic viruses. Most cancerous growths have disabled the interferon response and thus it does not usually present a major hurdle. On the other hand, effective interferon response in healthy cells surrounding tumors is essential for limiting their damage. Once oncolysis is in progress, the viruses released upon lysis of cancer cells are a target of the innate immunity, which has the potential to quickly destroy them, thus preventing infection of neighboring tumor cells. That possibility is addressed by combining delivery of oncolytic viruses with immunosuppressive drugs, buying them time to infect more cancer cells. Last but not least, the tumor microenvironment may interfere with the OV's potential to stimulate long-lasting antitumor immunity. Strategies addressing that problem were discussed earlier in this subsection.

9.2.4 Challenges and Limitations of Oncolytic Viruses

The concept of fighting cancer with OVs is truly fascinating from the perspective of basic science and hardly anyone would expect that when dealing with two very complex and continuously evolving phenomena, such as malignancy and viral infection, there will be no challenges. In addition to their efficacy, the clinical applications of OVs depend on straightforward methods for manufacturing, storage, and administration, as well as developing stringent safety protocols accounting for aspects specific to virotherapy. In general, it is very likely that these will benefit from experiences and knowledge stemming from current practices of vaccine manufacturing and management of infectious diseases. Developing detailed treatment protocols and means for training of all levels medical personnel are also a must. Among the big questions that need to be answered are: How does virotherapy fit in the current treatment protocols? What are the best ways to deliver OVs: individually or in combination with other approaches; intratumorally or systemically? When is virotherapy likely to be most effective: when every other available approach fails; concurrently with surgery or chemotherapy; before any other approach is attempted? It is very likely that even when we have answers to the above and other similar questions, they will be to some extent cancer-specific and OV-specific, bringing the realities of virotherapy along the lines of the never-ending discussion whether medicine is more like science or more like art. Treatments of inoperable tumors and chemotherapy-resistant tumors will most likely pave the way of OVs in regular clinical practice. Cancer biomarkers correlating malignancies and cell surface molecules could be of huge help in guiding virotherapy choices.

Currently, general contraindications for virotherapy are immunocompromised status and pregnancy; however, depending on the nature of the OV, more general contraindications may come to play. For example, it would be unreasonable to apply herpes-based virotherapy in an individual who is undergoing daily antiherpes therapy (acyclovir, valtrex, or similar) aiming to inactivate herpesviruses. So far, most clinical applications favor intratumoral OV delivery versus systemic delivery with the main concern being fast immune neutralization. Approaches for temporarily taming the immune system to allow sufficient persistence of intravenously injected viruses, OV serotype alternating, evolving virus surface molecules to be unrecognizable by neutralizing antibodies are being discussed as potential solutions of the problem. Viral particle persistence, targeted delivery, and particle neutralization are also obstacles in drug delivery with virus-based nanocontainers and recombinant vaccine development, both fields subject to active research and poised for growth.

Another source of concern is the possibility for virus transmission upon close contact with a patient undergoing virotherapy, both in clinical settings and at home. Technically, this should not be a huge issue since minimum pathogenicity is a critical requirement for OVs and practices for handling corresponding or similar infectious diseases are already in place. Most likely, the issue can be circumvented with additional training and clever protocol design. OVs incorporating molecular markers allowing for easy tracing can greatly contribute to a detailed understanding of virotherapy kinetics, as well as to resolve safety-related concerns and prevent unwarranted public health scares in the future.

Clinical trials of combinatorial therapy of OV and other anticancer approaches are being developed. Of special interest is the combination of inhibitors of immune system checkpoints and OVs. Drugs inhibiting immune system checkpoints (Opdivo, Keytruda, among others) block signals that control activation of T cells modulating inflammation and tolerance to self-antigens. The drugs, usually antibodies against key signaling components, are a relatively new tool in oncology and currently recommended for patients with inoperable tumors or advanced cancers failing to respond to more conventional treatments. Promising preclinical data are available for combinatorial protocols employing chemotherapeutical and OV approaches, as well as OV approaches and agents modulating immune function.

9.2.5 Oncolytic Viruses in Action

Identifying, designing, or selecting effective OVs is a rather daunting process and it comes as no surprise that OVs that are approved or in advance testing are based on long-researched viruses considered for other clinical applications, such as vaccines, and gene therapy, or viruses causing generally mild infections. The plethora of knowledge regarding virus biology, availability of research tools for addressing questions on cellular and organismal level, and availability of established protocols relevant to clinical applications are all factors facilitating the development of successful OV platforms. Table 9.1 lists key oncolytic viruses in clinical trials in late 2014/early 2015. The examples discussed below are on the forefront of the field and have already gained approval or are steadily moving torward one.

9.2.5.1 T-Vec

T-Vec is an armed oncolytic virus based on HSV-1, equipped to express human granulocyte-macrophage colony-stimulating factor (GM-CSF) inserted in the place of both copies of the RL1 gene coding for ICP34.5 (Infected Cell Protein 34.5). RL1 is a gene positioned in the repeated sequences flanking the UL (unique long) segment of the herpes genome, hence the virus has two copies. ICP 34.5 plays a central role in HSV-1 pathogenesis. It promotes neurovirulence, inhibits interferon-induced shutoff of protein synthesis allowing the viral infection to unfold, and interferes with autophagy. As a safety measure, the gene is deleted in all HSV-1 based oncolytic viruses in development at the cost of diminished replication capabilities. T-Vec also harbors a deletion of the US12 gene (the 12th gene found in the unique short segment of the herpes genome) coding for ICP47. The protein is instrumental in the inhibition of host immune response. It interacts with transporters associated with antigen processing and blocks the presentation of viral antigens in the context of MHC class I molecules. As a result, infected cells are masked for immune recognition by cytotoxic T-lymphocytes. GM-CSF is a growth factor secreted by several cell types from the innate immune system and functions as a cytokine stimulating the proliferation and differentiation of granulocytes, a class of white blood cells, mounting fast inflammatory response. T-Vec is currently recommended as a treatment of melanoma, although during the safety phase of testing, it has been tested against a broad array of cancers. It is administered via local injection directly in melanoma tumors on the skin surface, subcutaneously, or in lymph nodes. It propagates selectively in malignant cells that have mutations altering the integrated stress response system and protein synthesis shutoff mechanisms of the cell. It is unable to propagate in

TABLE 9.1 Examples of Oncolytic Viruses in Clinical Trials

| Virus | Manufacturer | Modification | Number of Clinical Trials | | | Cancers |
			Phase I	Phase II	Phase III	
ADENOVIRUS						
Onyx-015	Onyx Pharmaceuticals	Type 2/5 chimaera, E1B deletion	6	6	6	Head and neck cancer, pancreatic cancer, ovarian cancer, colorectal cancer, gliomas, lung metastases, and liver metastases
H101	Shanghai Sunwaybio	E1B deletion, partial E3 deletion	1	2	1	Squamous cell carcinoma and head and neck cancer
DNX-2401	DNAtrix	Δ24-RGD insertion	4	0	0	Glioblastoma, ovarian cancer
VCN-01	VCN Biosciences	PH20 hyaluronidase insertion	2	0	0	Pancreatic cancer
Colo-Ad1	PsiOxus Therapeutics	Chimeric Ad11/3 group B	1	2	0	Colon cancer, NSCLC, renal cancer, bladder cancer, and ovarian cancer
ProstAtak	Advantagene	TK insertion	4	1	1	Pancreatic cancer, lung cancer, breast cancer, mesothelioma, and prostate cancer
Oncos-102	Oncos Therapeutics	Δ24-RGD-GM-CSF insertion	1	0	0	Solid cancers
CG0070	Cold Genesys	GM-CSF and E3 deletion	1	1	1	Bladder cancer
VACCINIA VIRUS						
Pexa-vac (JX-594)	Jennerex Biotherapeutics	GM-CSF insertion, TK disruption	7	6	0	Melanoma, liver cancer, colorectal cancer, breast cancer, and hepatocellular carcinoma
GL-ONC1	Genelux	TK disruption, haemagglutin disruption, F14.5L disruption	4	1	0	Lung cancer, head and neck cancer, and mesothelioma

HERPESVIRUS						
T-VEC	Amgen	ICP34.5 deletion, US11 deletion, GM-CSF insertion	2	3	2	Melanoma, head and neck cancer, and pancreatic cancer
G207	Medigene	ICP34.5 deletion, UL39 disruption	3	0	0	Glioblastoma
HF10	Takara Bio	UL56 deletion, selected for single partial copy of UL52	2	1	0	Breast cancer, melanoma, and pancreatic cancer
SEPREHVIR(HSV1716)	Virttu Biologics	ICP34.5 deletion	5	1	0	Hepatocelluar carcinoma, glioblastoma, mesothelioma, neuroblastoma
OrienX010	OrienGene Biotechnology	ICP34.5 deletion, ICP47 deletion, GM-CSF insertion	1	0	0	Glioblastoma
REOVIRUS						
Reolysin	Oncolytics Biotech	None	15	9	0	Glioma, sarcomas, colorectal cancer, NSCLC, ovarian cancer, melanoma, pancreatic cancer, multiple myeloma, head and neck cancer
SENECA VALLEY VIRUS						
SVV-001	Neotropix	None	3	1	0	Neuroendocrine-featured tumours, neuroblastoma, and lung cancer
COXSACKIEVIRUS						
Cavatak (CVA21)	Viralytics	None	3	1	0	Melanoma, breast cancer, and prostate cancer

GM-CSF, granulocyte-macrophage colony-stimulating factor; NSCLC, non-small-cell lung cancer; RGD, Arg-Gly-Asp motif; TK, thymidine kinase; US11, unique short 11 glycoprotein.
Reprinted from Kaufman HL, Kohlhapp FJ, Zloza A. Oncolytic viruses: a new class of immunotherapy drugs. Nature Reviews Drug Discovery 2015;**14**(9):642–2. with permission from Nature Publishing group.

healthy cells due to the lack of the ICP 34.5 protein, which overrides the above defenses under normal circumstances. GM-CSF is expected to stimulate the innate immune response upregulating local inflammation and thus contributing to the tumor destruction. The latter has not been demonstrated very convincingly so far. T-Vec application in combination with other anticancer tools is in consideration. Efforts to balance the ICP34.5 impact on pathogenicity of HSV-1-based oncolytic viruses without attenuating their ability to replicate in cancer cells are underway. Targeted deletions of individual domains of ICP34.5 are aiming to separate its functions, in hope to identify mutants with high potency of killing malignant cells. The availability of an established and low-cost option of small molecule drug to control herpes replication (acyclovir, a chain terminator guanidine analog preferentially incorporated by the HSV DNA pol) is a huge plus in the development and applications of HSV-1-based oncolytic viruses. T-Vec OV has gained approval in the United States, Europe, and Australia for treatment of melanoma and it is currently in clinical trials for other cancers and as part of combinatorial approaches.

9.2.5.2 PVS-RIPO

PVS-RIPO is a poliovirus-based OV developed at Duke University, which gained the status of a breakthrough therapy for glioblastoma in 2016. PVS-RIPO is a genetically modified OV originating from the Sabin type 1 polio strain used for oral vaccination. Genetic engineering has replaced the poliovirus internal ribosomal entry site (IRES) with one from human rhinovirus type 2, abolishing neurovirulence. Polio is a (+)ssRNA virus, whose genome is translated immediately upon entry in the cell from a tissue-specific IRES. The IRES ensures fast and efficient translation by allowing ribosomes to start synthesizing viral proteins with minimal participation of translation initiation factors, speeding up the rate-limiting step of translation. The engineering of heterologous IRES preserves the ability of the virus to support viral protein synthesis and propagation; however, since the replacement IRES is not active in neuronal cells, the modified OV lacks its propensity to replicate in neurons. PVS-RIPO is administered intratumorally via catheter and can infect and lyse any cell expressing the CD155 molecule, which happens to be the poliovirus receptor. CD155 is commonly overexpressed in glioblastoma and promotes tumor cell migration, invasion, and metastasis. CD155 is also found on the surface of antigen-presenting cells; however, polio infection is known to not interfere with their ability to trigger activation of the immune system. At this moment, it is not well understood how much of the promising results seen in the clinic are connected to direct oncolysis and/or stimulation of the immune system.

9.3 VIRUSES AND GENE THERAPY

Gene therapy is a therapeutic approach of targeted delivery of nucleic acids to cells for the purpose of replacing, repairing, or regulating genes related to treatment or prophylaxis of diseases. Interest in gene therapy has been around for a long time as the idea of eliminating the cause of genetic diseases is very appealing and relevant to many people with diverse pathologies. Recent advances in the fields of gene therapy vectors and genetic manipulations of mammalian cells in vitro allowing genome editing has renewed the interest in gene therapy and elevated hopes for finding long-term drug-free solutions for genetic diseases. Clinical trials of gene therapy targeting cancer currently present the largest fraction of trials associated with the approach.

Technologically, gene therapy approaches can be applied in vivo or ex vivo (Fig. 9.6). The in vivo approaches aim to modify the genetic makeup of target cells, while they are residing in the body of the patient. The ex vivo approaches include isolation of target cells and their genetic manipulation in tissue culture, after which they are transferred back to the patient body. Both in vivo and ex vivo approaches target human somatic cells, thus the changes introduced by gene therapy and their benefits are not heritable and are only applicable to the individual undergoing treatment. Manipulations of germ line cells are technically possible; however, they are not being pursued for safety and ethical reasons. Pursuing gene therapy approaches is only feasible for genetic diseases whose molecular basis is well understood. Among them, diseases associated with single genes and single type of mutations have the greatest chance to benefit from gene therapy. For example, sickle cell anemia is a monogenic disease resulting from a single nucleotide mutation replacing the amino acid glutamate (charged and hydrophilic amino acid) with a valine (hydrophobic amino acid) in the molecule of the hemoglobin. Unfortunately, the substitution interferes with the structural integrity of the protein and causes aggregation, thus diminishing the hemoglobin oxygen–binding capacity. A gene therapy protocol for correcting the genetic defect responsible for sickle cell anemia would require replacement of one gene with average size or editing of one nucleotide and the protocol would be applicable to every single patient with the disease. Cystic fibrosis on the other hand, although a monogenic disease, is associated with a large number of heterogeneous mutations in a very large gene, thus requiring many different approaches or at least runs of the same technology after the specific genetic defect is identified by sequencing. The different biology of blood cells and lung epithelial cells also presents an obstacle for successful gene therapy. Gene therapy approaches are especially appealing for genetic diseases associated with short life span, complicated treatment regiments, and poor quality of life. Understandably, many gene therapy–related efforts are focused

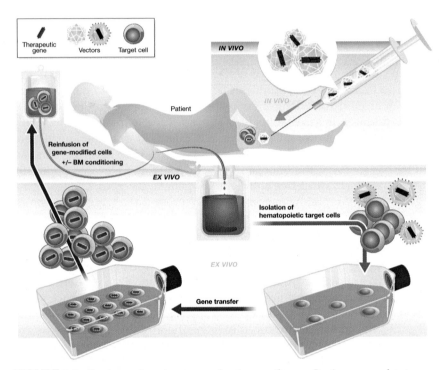

FIGURE 9.6 **In vivo and ex vivo approaches to gene therapy.** In vivo approaches to gene therapy involve delivery of gene therapy vectors directly in the body of the patient, most frequently in the targeted organ. Ex vivo approaches involve isolation of patient cells and their manipulation in laboratory settings to achieve the necessary genetic makeup. Subsequently engineered cells are reintroduced in the patient and expected to restore normal physiology. Depending on the specifics of the procedure, bone marrow conditioning may be required. *Image courtesy Kaufmann, KB, Büning H, Galy A, Schambach A, Grez M. Gene therapy on the move.* EMBO Molecular Medicine *2013;5(11):1642–61.* https://doi.org/10.1002/emmm.201202287, *CCBY 3.0,* https://creativecommons.org/licenses/by/3.0/.

on children with genetic diseases, as well as on cancer therapies. Armed oncolytic viruses can also be considered a form of gene therapy.

Recently, progress has been made in approaches applied to hematopoietic stem cells (HSCs) and T cells ex vivo as well as liver-directed in vivo gene therapy; however, active clinical trials in several other areas are also undergoing. Mechanistically, gene therapy approaches aim to introduce genomic modifications or to regulate gene expression (Fig. 9.7). Genomic modifications include replacement of the defective gene leading to recovery of the wild-type protein in the affected cells and restoration of the wild-type function. Gene insertions can lead to introduction of a new protein exerting therapeutic function. Both approaches were repeatedly attempted over the years with limited success; however, recent developments in relevant viral vectors have delivered promising results.

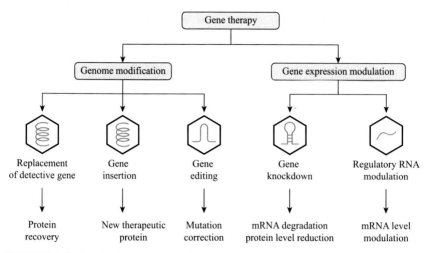

FIGURE 9.7 **Mechanistic aspects of gene therapy.** Gene therapy interventions aim to modify the genome to correct genetic defects and restore functionality by replacing defective gene copies, inserting new genes, or editing the specific mutations (left). Alternatively, gene therapy can also target the level of gene expression (right). The cellular concentration of specific proteins can be modulated by targeting their cognate mRNAs for degradation or influencing gene expression by altering the expression of regulatory microRNAs.

Gene editing has become possible with the boom of gene editing technologies such as Transcription activator–like effector nuclease (TALENs), zinc finger nucleases (ZFN), and the CRISPR/Cas system (Chapter 3). Gene expression modulation is targeted mainly utilizing RNAi technology applied directly to relevant mRNAs or indirectly to regulatory RNA molecules. In addition, tissue-specific promoters and promoters allowing regulated expression are being used as tools to modulate gene expression in the target cell and/or the expression of the introduced transgene.

Traditionally, the following groups of viruses have been explored as gene therapy vectors: retroviruses (including Lentiviruses), adenoviruses, adeno-associated viruses, herpesviruses, and baculoviruses. Brief overview of properties relevant to gene therapy is provided in Table 9.2. The evolution of some of the viral vectors will be described in the examples of success stories of ex vivo and in vivo gene therapy.

9.3.1 Ex Vivo Gene Therapy

Ex vivo gene therapy approaches are most frequently applied to hematopoietic stem cells (HSCs), which are relevant to blood and immunological disorders, cancer, and genetic diseases that affect tissues and organs easily accessible by blood cells. Among diseases potentially treatable with genetically modified HSCs is severe combined immunodeficiency (SCID, popularly known as bubble boy disease), whose history of treatment can

TABLE 9.2 Key Properties of Viral Vectors for Gene Therapy

Vector	Genome/Capacity	Advantages	Limitations
Retrovirus	ssRNA/8 kb	• Theoretically stable transgene expression • Low immunogenicity • No (or very low) preexisting immunity	• Transduce only dividing cells • Possible insertional mutagenesis • Limited insert size of the transgene • Handled at BSL-2 facilities
Lentivirus	ssRNA/8 kb	• Transduce nondividing and dividing cells • Stable transgene expression • Absence of preexisting immunity	• Possible insertional mutagenesis • Limited insert size of the transgene • Handled at BSL-2 facilities
Adenovirus	dsDNA, 36 kb	• Transduce nondividing and dividing cells • High-level transgene expression • Vector particles produced at high titers	• Transient transgene expression • Broad preexisting immunity • High immunogenicity
Adeno-associated virus	ssDNA, <5 kb	• Transduce nondividing and dividing cells • Nonpathogenic with wide cellular tropism • Capable of long-term transgene expression • Low immunogenicity	• Preexisting immunity • Limited insert size • Possible transgene integration
Baculovirus	dsDNA/~130 kb	• Nonpathogenic to humans • Lack of integration • No preexisting immunity • Large cloning capacity	• Transient transgene expression • Inactivated by serum complement • limited long-term stability when passaged

Modified from Sung L.-Y, Chen C.-L, Lin S.-Y, Li K.-C, Yeh C.-L, Chen G.-Y, Lin C.-Y, Hu Y.-C. Efficient gene delivery into cell lines and stem cells using baculovirus. Nature Protocols 2014;9(8):1882–99, and reprinted with permission from Nature Publishing group.

somewhat be viewed as the history of gene therapy itself. Children with SCID exhibit severe immunodeficiency and are prone to infections due to inability to execute proper immune response. If not treated, children with the disease rarely survive past 2 years of age. Traditional treatment involves bone marrow transplantation early in life replacing the hematopoietic cells harboring mutant genes with matching cells carrying healthy genes. Two major types of SCID, X-linked SCID and ADA (adenosine deaminase)-SCID, have been the subject of gene therapy. X-linked SCID develops in patients with mutations in the γ_c chain shared between multiple interleukin receptors on the surface of B and T lymphocytes. Nonfunctional γ_c chain prevents proper interleukin signaling between immune cells and halts their development, thus effectively abolishing the function of the immune system. ADA-SCID develops as a result of ADA deficiency leading to defects in purine degradation and unbalanced nucleotide pools that arrest cell proliferation. Defects in the immune system are observed due to low rate of lymphocyte proliferation. Early gene therapy experiments utilized murine viral vectors and were largely unsuccessful, presumably due to poor retention of the genetically modified HSCs. First reports about successful gene therapy trials were announced in 2000 using ex vivo transfer of γ_c cDNA encapsulated in Moloney murine leukemia retrovirus into autologous HSCs of patients with X-SCIDs. The generated excitement in the medical and scientific community was quickly extinguished by the development of leukemia in several patients in two different clinical centers in Europe. Clinical trials were cancelled and concerns about safety of gene therapy were reinforced. Molecular analysis revealed that the retroviral vector has integrated near an oncogene, LMO2. It was suggested that the transcriptional enhancers located in the long terminal repeats of the vector may have upregulated cellular proto-oncogenes adjacent to the insertion site, thus promoting insertional oncogenesis. In addition, it was established that some of the patients who developed leukemia already had mutations in LMO2. For the following decade, scientists focused their efforts on creating self-inactivating (SIN) retroviral and lentiviral vectors. SIN vectors lack the enhancer sequences normally found in the long terminal repeats, thus minimizing the chance of insertional mutagenesis, but preventing the incorporated exogenous gene from expression. To ensure that the latter is expressed, a promoter is inserted just before the sequence of the gene of interest. Experiments showed that the SIN retroviral vectors support the expression of the therapeutic gene and HSCs genetically modified using such vectors are well tolerated. The approach was used to generate SIN lentiviral vectors, which have the capability to infect nondividing cells, allowing for simpler protocols for genetic modifications. In 2016, the European Medicine Agency approved Stremvelis, a gene therapy approach to correct ADA-SCID in children for whom no matched HSC transplantation donor is available (~80% of the cases). The ups and downs of SCID gene therapy development have provided multiple useful data sets over a period spanning two decades and continue to do so. The technology is now being applied toward blood disorders (sickle cell anemia, thalassemia) using vectors driving strictly controlled expression of healthy hemoglobin; neurological and lysosomal storage diseases using genetically modified HSCs as portable protein factories; creating anticancer T cells with T-cell receptors engineered to recognize specific cancer antigens; and creating ccr5 deletions to confer HIV resistance.

9.3.2 In Vivo Gene Therapy

The preferred targeted organ for in vivo gene therapy is the liver. The liver is built from long-living hepatocytes that are metabolically active and can easily receive and release substances and particles through the bloodstream. Hepatocytes easily synthesize large amounts of proteins and secrete them in the blood, thus offering possibilities for correction of blood and metabolic disorders not directly associated with the organ; for example, hemophilia and other blood clotting disorders. Hemophilia patients are missing functional Factor IX, preventing blood clotting and thus are at high risk for detrimental accidental hemorrhages. Current therapies include Factor IX replacement therapy, which is delivered intravenously. The therapy is very costly and only available in countries with advanced health care systems. Interestingly, it has been calculated that the presence of 1% of the normal levels of Factor IX can reduce the risk of accidental hemorrhages significantly and reduce the need of the costly and inconvenient prophylactic measure. Adeno-associated virus (AAV) was found to have natural tropism to liver cells and to remain stably associated with transduced cells persisting as an episome in the nucleus. When injected intravascularly in animal models, AAV did not trigger immune response; instead, it promoted transgene-specific tolerance. Despite that, initial trials demonstrated only transient factor IX expression. Subsequent improvements of the gene therapy construct and the identification of AAV serotype 8 (AAV8) with enhanced liver tropism allowed for 6% expression of the normal Factor IX levels, which effectively allowed patients to significantly reduce or even eliminate the need of prophylactic Factor IX infusions, thus not only eliminating significant health risk but also improving quality of life. Furthermore, expression was steady and stable at the 3-year checkpoint post–gene therapy application with no major side effects. Some of the patients received immunosuppressive steroid treatment to avoid liver damage due to inflammation 4–8 weeks post–gene therapy vector application. AAV-mediated gene therapy has shown promise in skeletal and heart muscle injections, as well as subretinal injection for vision correction. The general consensus is that the AAV platform offers opportunities for safe and reasonably efficient gene therapy, which could be further improved if the dosage of the virus particles is lowered. Efforts are being focused on modifying and/or evolving AAV capsids to increase selective affinity for different tissues, with the goal to lower the risk of accidental transduction in nontargeted cell types.

Muscle tissues are also being actively explored as a destination for delivery of virus-based gene therapy preparation. Glybera was the first gene therapy drug approved in the Western world (European Union) in 2012, delivering a copy of functional lipoprotein lipase (LPL) gene intramuscularly to patients with LPL deficiency. Although the drug is available, getting it to the patients in need is prohibitive due to high cost, mainly driven by the rare occurrence of the disease. Intramuscular delivery of genes coding for broadly neutralizing antibodies against HIV and influenza has shown promise in mice injected with AAV1 vector and has been proposed to be considered as a possible alternative to vaccination. Attempts to improve heart vascularization by injecting genes for vascular growth factors have not delivered meaningful results so far. In vivo gene therapy has been attempted in lungs in the context of cystic fibrosis; however, the mucus of the diseased lungs appears to be a huge barrier for gene delivery. Progress has been made in addressing some retinal diseases; however, the low frequency of occurrence of these diseases and the high price tag of the approach are current hurdles for further development.

9.3.3 Gene Therapy Perspectives

Recent advances in gene therapy renewed the interest in the approach and brought it back in the realm of technologies worth investing in. As with any new health technology with a complex nature, concerns and challenges are coming together with promising developments. Natural immune response to viral vectors remains a major challenge for gene therapy in vivo. At least in part, its neutralization effect can be alleviated with proper selection of vectors with lower immunogenicity, modifying vector surface molecules, or applying gene therapy in combination with immunosuppressive drugs to "buy time" for efficient gene transfer. Insertional mutagenesis continues to be on the "watch list" of possible challenges while long-term data are being collected from clinical trials with the newest generation of retroviral and lentiviral vectors. Vector bioselection and surface engineering are expected to improve vector targeting, thus potentially increasing the efficiency of transfer, lowering the dose of the administered vectors (and thus cost), and minimizing the risk of gene delivery to cell types different from the intended ones. Baculoviruses are emerging as a new promising gene therapy platform. They have been shown to naturally enter mammalian cells, which cannot sustain their propagation due to promoter incompatibility. When genes under the control of mammalian promoters were delivered with baculoviruses, adequate protein expression was executed (BacMam technology). Another great advantage of baculoviruses is their large coding capacity, which exceeds more than 10 times the coding capacities of retroviral, lentiviral and AAV-based gene therapy vectors and more than three times the coding capacity of adenoviral vectors (Table 9.1). Large coding capacity allows more room for genetic manipulations related to expression regulation, incorporation of genetic markers for tracing the whereabouts of the gene therapy vectors, as well as addressing conditions that are driven by genetic defects in multiple genes or in multicomponent complexes. Recently, a new modular multigene delivery system based on baculoviruses was developed (MultiPrime) allowing multigene delivery and expression in both mammalian and insect cells. Proof of principle experiments demonstrated efficient transduction of many types of mammalian cells including nondividing primary neurons and induced pluripotent stem cells (iPSCs), as well as the ability to support gene delivery for the purpose of CRISPR/Cas 9 gene editing. Although production and applications of baculoviruses are not flawless (for example, they are unstable when passaged long-term and are prone to inactivation by the complement), they have been studied and engineered for a very long time as tools in protein expression and biocontrol (Chapters 4 and 7, respectively), thus many issues relevant to safety and manufacturing are solved, are being addressed, or have been on the radar of the scientific community for years. Looking at the evolution of the technology for generation of recombinant baculoviruses for protein expression purposes (discussed in Chapter 4), one can appreciate the body of knowledge and strategies for improvement that baculoviruses are potentially bringing to the field of gene therapy. In addition, baculoviruses are being developed into a surface display platform similar to phage display and are also considered a promising tool in virus-based nanotechnologies, thus the potential for fast advancements is high. Baculoviruses are also currently used to physically package some of the viral vectors used for gene therapy.

Another field that is expected to influence gene therapy rather dramatically is iPSC cell reprogramming, which will allow a broad set of cell types to be manipulated ex vivo for the purpose of autologous transplantation (Fig. 9.8). It is envisioned that when iPSC reprogramming is developed into regular practice, patient-derived somatic cells will be reprogrammed in iPSCs, which will be subjected to gene editing using ZFNs, TALENs, or CRISPR/Cas approaches to correct the existing genetic defect. The iPSC with a corrected defect will be then differentiated into the cell type of interest and transplanted in the body of the patient. Although it is quite possible that it will take a while to develop an inventory of protocols for generating a substantial number of cell types, the general principles of the approach are currently being tested in the ex vivo development of T cells harboring the ccr5 deletion associated with resistance to HIV.

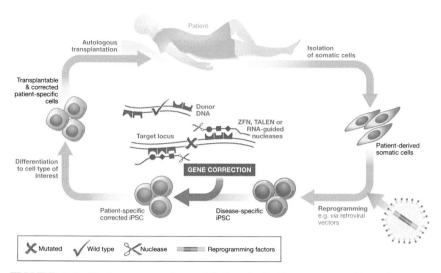

FIGURE 9.8 **Designer corrections of induced pluripotent stem cells with genetic defects for the purpose of autologous transplantation.** Patient-specific induced pluripotent stem cells (iPSCs) derived by reprogramming somatic cells are subjected to genetic corrections employing Zinc finger nucleases (ZFN), TALEN (Transcription activator–like effector nuclease), or RNA-guided nucleases (CRISPR/Cas system) and reintroduced in patient's body via autologous transplantation. The procedure uses viruses to deliver key molecules for reprogramming, as well as templates and tools for editing. *Image courtesy Kaufmann KB, Büning H, Galy A, Schambach A, Grez M. Gene therapy on the move.* EMBO Molecular Medicine *2013;5(11):1642–61.* https://doi.org/10.1002/emmm.201202287, *CCBY 3.0,* https://creative-commons.org/licenses/by/3.0/.

Phage therapy, oncolytic viruses, and virus-based gene therapy share many common features. Intellectually, the principal ideas underlining each one of them came to light many years before the necessary knowledge and tools were available to demonstrate practical benefits in safe, economically, and socially acceptable ways. All of them went and most likely will go in the future through ups and downs gaining strength from successes and losing ground from setbacks. All of them have benefited tremendously from the advances of various scientific disciplines, especially molecular biology, and most likely will require significant shifts in current health-related thinking before they become established. Until now, most of the medical interventions were in the realm of controlled interaction between the diseased human and more or less static medical tools, continuously improving over time. Harnessing the power of viruses and applying it to medical technology in 21st century is more along the lines of interaction between two systems evolving in real time and thus calls for changes elsewhere to account for that. Dealing with complex phenomena is challenging in any context; however, as long as the society and science keep looking for solutions and answers, chances are that such will be found sooner or later.

It is hard to predict whether the beneficial impact of viruses to our society will become widely recognized and appreciated any time soon; however, it is certain that there are many exciting developments to come.

Further Reading

1. Collins M, Thrasher A. Gene therapy: progress and predictions. *Proc R Soc B Biol Sci* 2015;**282**(1821).
2. Cooper CJ, Khan Mirzaei M, Nilsson AS. Adapting drug approval pathways for bacterio-phage-based therapeutics. *Front Microbiol* 2016;**7**:1209.
3. Chiocca EA, Rabkin SD. Oncolytic viruses and their application to cancer immunother-apy. *Cancer Immuno Res* 2014;**2**(4):295–300.
4. Kaufman HL, Kohlhapp FJ, Zloza A. Oncolytic viruses: a new class of immunotherapy drugs. *Nat Rev Drug Discov* 2015;**14**(9):642–62.
5. Lawler SE, Speranza MC, Cho CF, Chiocca EA. Oncolytic viruses in cancer treatment: a review. *JAMA Oncol* 2017;**3**(6):841–9.
6. Mansouri M, Bellon-Echeverria I, Rizk A, Ehsaei Z, Cianciolo Cosentino C, Silva CS, Xie Y, Boyce FM, Davis MW, Neuhauss SC, Taylor V, Ballmer-Hofer K, Berger I, Berger P. Highly efficient baculovirus-mediated multigene delivery in primary cells. *Nat Commun* 2016;**7**:11529.
7. Miedzybrodzki R, Borysowski J, Weber-Dabrowska B, Fortuna W, Letkiewicz S, Szufnarowski K, Pawelczyk Z, Rogoz P, Klak M, Wojtasik E, Gorski A. Clinical aspects of phage therapy. *Adv Virus Res* 2012;**83**:73–121.
8. Naldini L. Gene therapy returns to centre stage. *Nature* 2015;**526**(7573):351–60.
9. Ungerechts G, Bossow S, Leuchs B, Holm PS, Rommelaere J, Coffey M, Coffin R, Bell J, Nettelbeck DM. Moving oncolytic viruses into the clinic: clinical-grade production, puri-fication, and characterization of diverse oncolytic viruses. *Mol Ther Methods Clin Dev* 2016;**3**:16018.
10. Wittebole X, De Roock S, Opal SM. A historical overview of bacteriophage therapy as an alternative to antibiotics for the treatment of bacterial pathogens. *Virulence* 2014;**5**(1):226–35.

Index

Note: 'Page numbers followed by "f" indicate figures and "t" indicate tables.'

Printed in the United States
By Bookmasters